PROFESSIONAL

I0071638

CITIZEN MANUAL 9
ADVERSE CONDITIONS
And
ENVIRONMENTS
(ACE)

END USER WARNING: All information in this publication is solely for entertainment and informational purposes only; this work does not contain classified or sensitive information restricted from public release. The authors and publisher accept no liability or responsibility for application, use, or misuse of the tactics, techniques, and procedures contained in this work or any injury or loss through use of the content of this publication.

Developed for

The Professional Citizen Project

www.TPCproject.com

This book is dedicated to my amazing wife and children who have helped encourage me, and have been supportive, of my love for the outdoors! They push and inspire me to be better every day!

Citizen Manual (CM-9)
ADVERSE CONDITIONS and
ENVIRONMENTS (ACE)

Copyright © 2023 Jay Pallardy

All rights reserved. All rights to non-governmental content are protected by copyright, no non-governmental part of this book may be reproduced or transmitted in any form or by any means without written permission from the author.

The appearance of U.S. Department of Defense (DoD) visual information does not imply or constitute DoD endorsement.

ISBN   979-8-9895092-0-1

Printed in the USA by The Professional Citizen Project

www.TPCproject.com

CITIZEN MANUAL 9
ADVERSE CONDITIONS and ENVIRONMENTS

INTRODUCTION

Why do we need a Professional Citizen manual for outdoor gear and survival skills?

There is a ton of information out there on wilderness survival and related gear and, thanks to social media, things like Bushcraft and the Minuteman concept have become more popular than ever. But sorting through this vast amount of information and trying to decide what you need to know and why can be problematic, especially for anyone not accustomed to living and working outdoors. Getting gear is the easy part, but learning how to use it in adverse conditions is the hard part. A big part of this is how do we take this information, make it relevant to you, the Professional Citizen, and make sure you know how the required skills, gear, and concepts can be used to your advantage in the context of being The Professional Citizen! I like to think of it as "Tactical Bushcraft" where we are more Frontiersman or Mountain Man in our approach to these skillsets and less Operator-like.

So, in this reference manual, we will discuss not only the concepts, skill sets, and gear needed to become better prepared to survive in the outdoors, but also the mindset required to be able to improvise, adapt, and overcome harsh conditions and environments.

Foreword

In the summer of 1995, I was lucky to go on a backpacking trip to Philmont Scout Ranch in New Mexico. This trip ignited a fire inside of me for a love of the outdoors that ultimately changed the trajectory of my life. I had camped for most of my young life, but to be immersed in the mountains was a completely different experience, almost a church-like experience that left me yearning for more of that adventure.

One year later I found myself invited to join a small private group on an unguided climb of Mount Rainier. I had no idea what I was getting myself into but decided to go for it anyways. Before the trip I took a small course on the basics of glacial travel, self-rescue, and working as a rope team. I was young, in shape, had all this brand new (and expensive) gear and I loved it. The adventure, the unknown dangers, the hard work, and it was great! But then that storm hit descending from the summit. White out conditions, a bit of panic, and an added sense of urgency to get to safety. I took lead and we successfully descended the mountain on pure compass bearings, a will to succeed, and a little bit of luck. We were exhausted, wet, cold, and hungry. We had been on the go for almost 30 hours at that point and you know what, I loved every minute of it!

I was lucky to get to go on a lifetime worth of adventures throughout the years and there was always a reoccurring theme amongst the folks that I went with and that was a combination of having a positive attitude, being masters of their craft, and being very giving of their experience to help me become better at my craft. This is where this book comes in. I have been able to reap the benefits of having mastered the basics and beyond and this is what has enabled me to enjoy almost 3 decades of adventures in the outdoors. I may not be the fastest nor the strongest anymore, but I probably enjoy it more than most and that is what I wish to impress upon you. I want you to have a greater understanding of the basics so that you too can be better prepared for whatever life may throw your way.

A much younger me ready to tackle the world.

When I was asked to write this book, I had a bit of fear of doing something I had never done before (you will see more on fear later on in this book) but I decided that it needed to be done. To share what I know with you. To take my own individual experiences in the outdoor community and combine it with best practices from both the climbing and military communities and put it together in a way that is useful to you, The Professional Citizen.

What is the Professional Citizen Reference Series?

The Professional Citizen series of references is a clearly written, easy to use set of references that address baseline individual tasks through complex small unit tactics - all done from the Citizen perspective. This set of references is not an Army or USMC manual; these publications are written to be Citizen specific. We have adapted relevant tactics, doctrine, and best practices for the Citizen and packaged them into a practical series of references. The Professional Citizen series references are written by subject matter experts from a wide variety of backgrounds including military, outdoorsmen, and industry professionals with decades of combined experience.

The Professional Citizen must train and execute in a resource constrained environment. Our community uses US military tactics as the starting point because they are proven and are the most available resource for us to reference. The challenge with using these military sources for our community is the system is designed for a non-resource constrained organization (the military). The methods and the resources far outstretch what we as Pro Citizens can muster. This is the challenge with current materials - finding relevant tactical and doctrinal content designed *specifically for the "average" Citizen* was nearly impossible. The Professional Citizen series of manuals does not include how to conduct air assaults, load helicopters, or emplace cratering charges. The CM manuals provide things that are relevant to you; referencing only those tasks directly related to anticipated Citizen missions.

This series of manuals is grounded in proven Tactics, Techniques, and Procedures (TTPs) and doctrine. This series of references is not a renamed Army manual with a new cover or a repackaged compilation of existing manuals. The references provide you, the Pro Citizen, with **applicable** knowledge and skill development. Yes, some content is a direct lift from current doctrine because a

4

particular subject or reference does not require adjustment to meet your requirements. These references are designed to provide the technical and tactical data in a manner that enables you to select the material that best suits your requirements. Everything from basic individual tasks through planning and leading organizations during a crisis is contained in the series.

Crawl, Walk, Run

The *"crawl, walk, run"* methodology for developing a fighting man (and leaders) is a simple but proven method. The crawl phase can apply to all members of the community no matter how experienced.

We are all new at everything at one point in our journey. Don't allow yourself to be gatekept or tacti-shamed into not improving...or even starting.

Building initial skills and proficiency is critical before moving on to more complex tasks. Learning a new foundational task or a set of tasks sets the bedrock to build on. You will start to add more complex tasks that these basic tasks support (referred to as nesting). As these come together you will build proficiency and credibility as a team member. Be realistic, be deliberate, and be patient with yourself as you build expertise. These references are organized so you can choose your entry point and select the skills that *you* require - from basic to complex. Skill levels vary among us, we all enter the process at different points on the spectrum. Building skills through this deliberate approach will ensure you have not missed a foundational / fundamental element of your training along the way.

This Manual, CM-9, Adverse Conditions and Environments (ACE)

The audience for this series of references is the Professional Citizen with a commitment to excellence who needs a framework to help prepare and train. CM-9 is an introduction to understanding the basic skills, philosophies, and equipment needed for surviving in the outdoors. Most of what you will read here contains proven concepts and basic tactical and technical skills that will be the bedrock to build upon as you work to master your craft. Practicing and mastering the skills in this reference builds a foundation for you to continue to develop and train more complex skillsets or can assist in sharpening your approach as a seasoned tactician or prepper. Understand that this manual is just a starting point that provides some of the necessary foundational competences to build upon with the skills and tasks contained in the other Professional Citizen series manuals and guides.

THE PROFESSIONAL CITIZEN PROJECT

If not you, who will?

CHAPTER 1
Outdoor Fundamentals

Why do we need to know the basics?

In today's environment, a lot of folks start off emulating what they see, which in many cases is a polished product of years and years of hard work. So many times, we, as everyday "Professional Citizens", don't get to see the backstory or processes that it took others to get to that point. The understanding of gear and its uses is a starting point to be able to move forward. This is where that "Crawl, Walk, Run" methodology comes into play. The basics help in creating that base foundation of skills to build upon as we progress in our overall level of preparedness. Experience acts as a shield against disaster, and it is our goal to set you off on this journey on the right foot!

As part of our desire to learn to survive in the outdoors, we need to learn the skills for safe and enjoyable existence in the backcountry. Not just for our own sake but for the sake of our families and friends we may decide to travel with. It is our duty to learn all the tools and techniques required for camping, navigation, first aid, survival, rescue, leadership, and so much more. I hope this book gives you a good start in this learning process.

Survival is a demanding activity, physically and emotionally. Most of us will never be world class athletes or operators, and it isn't necessary to become one. But the level of skill levels is rising amongst us in the civilian world. Good physical conditioning and mental fortitude are becoming increasingly important to give us an edge in the world of survival. Not only does it help you when times get tough, but it makes being outdoors more enjoyable now. With all that is going on in the world today, this lifestyle is proving to be more and more beneficial to just everyday living. We need to be positive, realistic, and honest with ourselves as to how we approach all of this.

In this book, we are going to cover gear basics, skills you should learn, share relatable experiences, and just plain old encouragement to getting on the right track. I hope you enjoy going through it as much as I enjoyed putting it together for you!

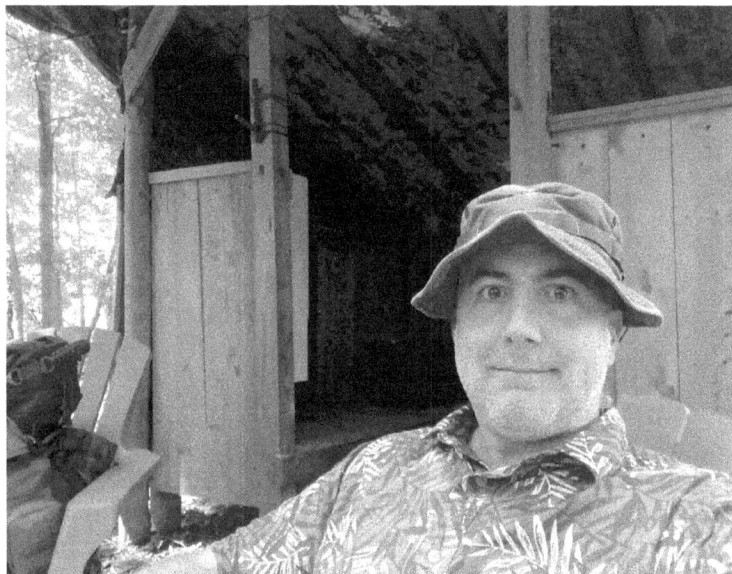

The author hanging out at his Bush Camp.

The 10 Essentials:

The special items that the Professional Citizen should always have in their kit has become known as the "Ten Essentials" This was put together many years ago in the mountain community as a way of identifying 10 important items to always have with you (or on you). There have been many variations of this list over the years and here we will identify and describe each one in more detail.

1. Map/Compass
2. Headlamp/Flashlight
3. First Aid
4. Water
5. Knife
6. Food
7. Extra Clothing
8. Rain Gear/Shelter
9. Fire Starter/Ignition Source
10. Cordage

1. Map/Compass

Not all who wander are lost, right? It's been a popular saying for years for outdoor enthusiasts around the country. But have you ever been lost? I mean really lost as in you have no idea where you are. I got lost for a day in the Uinta Mountain Range in Northern Utah on a solo backpacking trip. Wasted an entire day because I thought I knew where I was headed after breaking camp and didn't bother to check myself first with my map and compass.

Map and compass use is one of the most overlooked skill sets and a perishable one as well if not regularly practiced. Today we have GPS, cell phones, downloadable maps, etc but those don't mean anything if you don't know the basics of navigation. How to orient a map, triangulate to find your position, plot a course, use waypoints, take a bearing, shoot an azimuth, pace count, read a map, and more. These are the basics that every outdoorsman and Professional Citizen should know and practice with. GPS batteries can die, cell phones lose service and die, downloadable maps rely on batteries that, well die... You see the trend when you depend on technology to make up for lack of basic skills....

So, what should a basic navigation kit consist of? For everyone's daypack a compass with rotating bezel and clear base plate and a map of the area they are in. The compass should have a lanyard for keeping around their neck and the map should be waterproofed by laminating, in a map case, or in a zip lock bag.

What are some additional tools to go with your map and compass?

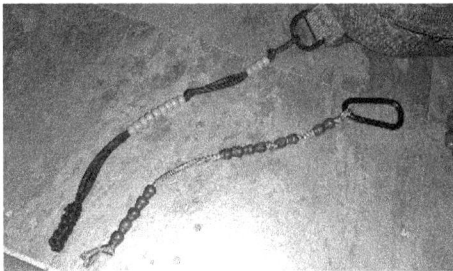

*Ranger Beads or pace beads for measuring the distance traveled
*Small write in the rain notebook and writing utensil for writing down bearings, distances, rotating landmarks/places of interest, leaving notes for others, etc
*Red light for navigating at night. Red light helps with maintaining good night vision and helps contour lines pop on a map in the dark. A white light can wash them out.
*Ruler/protractor, or string for measuring distances, marking grid coordinates, or for plotting course on a map. Most compasses also have a measuring scale on the sides that can also be used for this purpose.

Navigation can be a deep rabbit hole to go down, but the basics are simple, and everyone should know them. Technology is nice as an aid, but nothing beats knowing the fundamentals first.

2. Headlamp/Flashlight

Even if you have no intention of being out after the sun goes down it is essential to carry a light source. Light is essential for many reasons. It helps us find our way at night, helps with nighttime camp activities, can be used to aid in navigation, signaling for help or rescue, light can also be used as a friend/foe identifier.

Examples of light sources

Light can come from many sources including flashlights, headlamps, glow sticks, strobe lights, fire, and more.

-Headlamps: hands free light makes hiking and camp chores easier by freeing up your hands. They usually have an adjustable elastic head band and a tilt housing for the light itself.

-Handheld Flashlights: by far the most common lights out there. From tiny keychain lights to large work lights. Many handhelds have a pocket clip for carrying in your pocket and have either a tail cap button or side button.

-Hat Clip Lights: these lights are like a small handheld but with a clip for attaching to a baseball bat bill.

-Glow sticks/Chem lights: great secondary light source for area marking, tents, emergency lighting, making buzzsaws for signaling, and more. Hang one on the latrine at night for easy identification. Use different colors for playing capture the flag at night or for orienteering courses.

-Weapon lights: I'm a believer that every rifle should have a light on it (a sling too). Depending on your grip and setup you can either have a pressure pad switch or a press button setup. Learn how to properly use a weapons light.

-Rescue Strobes/Beacons: these specialty lights have been used by backcountry users and the military for years. They can be seen miles away and are designed to run for multiple days. They are also useful for nighttime bike rides for safety. Some models will flash SOS and others have an IR (infrared) feature for use under night vision devices.

The last point of consideration is the power source for your light. Standard batteries are the most common, but you need to bring spare batteries. Rechargeable lights are becoming more common but still require a power source to charge from which means bringing along some sort of battery bank to plug into.

Lights with a red lens option help with preserving your natural night vision and, in some uses, can keep your light signature down when you aren't wanting to be found. Don't forget spare batteries or charging options when choosing a light.

Finally, learn proper light discipline and practice it.

3. First Aid

The very nature of working in the outdoors, especially for the Professional Citizen, operating under distress, with heavy loads, various environments, working with sharp objects and things that create holes, makes it essential to have a first aid kit and the training to use it. Think about the activities we undertake, and the inherent dangers associated with them.

First Aid is one of the most overlooked of all the essentials. Folks like talking about all the gear like tents, packs, stoves, and more but when the discussion turns to First Aid then the conversation starts to fade away. I think it's because of the lack of training in the community. We are all always one step, one swing of an axe, one slip of a pocketknife, one spill of boiling water, one mishap around a firepit, one accidental discharge on the range from a bad accident but most people's First Aid kits are severely lacking to treat an injury beyond a scrape or booboo and worst yet is their training is at a level to put a Band-Aid on. These are hard truths, but we really need to look at First Aid in a more serious light.

First off, **GET TRAINING**!!! Red Cross 1st aid and CPR are not enough for what we do in the woods. At a minimum take a Wilderness First Aid class. Make sure it is a class that conducts scenario-based training in the field and makes you use your own kit. Take a Stop the Bleed course. This isn't just for an active shooter situation but also for major hemorrhaging from axe and knife injuries. These 2 classes should be minimum training for everyone. If you want to go beyond then take a Wilderness First Responder course, EMT basic training, or take a TECC (civilian equivalent of the military Tactical Combat Casualty Care course).

Secondly, your First Aid kit should represent the level of care you are trained to administer. Chest seals, tourniquets, and pressure bandages are just stuff if you don't know how to properly use them. Most common injuries are scrapes and bruises that are easily treated with a Band-Aid or two and some antibiotic ointment. Some injuries like sprains may require an Ace bandage. Think of the environmental factors as well when putting together your First Aid kit. Think of putting a package of antacids, like Tums, for upset stomachs and oral hydration salts in cases of dehydration. Pro tip: rolled gauze and tape will do wonders and cover most bleeding injuries, pack lots of rolled gauze.

First Aid kit examples

-Booboo kits... I'm a big fan of a simple booboo kit that should be in everyone's pack. A booboo kit is exactly as it says, it treats booboos. Bandaids, moleskin, antibiotic ointment, and tape about covers it. They fit easily in a cargo pocket so it's always on your person. Make them at get togethers so that everyone has their own!

-Trauma kits... This is separate from a "Booboo" kit in that it has the necessary items for dealing with massive hemorrhaging, airways, and broken bones. Tourniquets, Nasopharyngeal Airway Device (NPA), pressure dressing like Israeli bandages, SAM Splints, and so on and so forth are necessary for trauma in the field. If you are in the backcountry and hours (maybe days) away from that next level of care you, as a leader, need to have the ability to treat and stabilize that injury while you either wait for help or you are in a situation where you need to prepare to move the injured to an area that is safe and accessible to rescuers. This is big and why it is so important to seek the advanced training mentioned above.

Other considerations are things like TCCC casualty cards, trauma shears, markers, light sticks, wet wipes, nitrile gloves, space blankets, and more.

I could really go on and on, but I think we get the point on the importance of a proper First Aid kit and being trained in its use. If you haven't had any training, go get it now.

More details on first aid will be in Chapter 2.

4. Water
Water is one of the top 3 survival priorities so it should come as no surprise to see water on this list. The means to carry water and procure/treat it are always essential.

Good old high-quality H2O. You need water, without it you will die. Plain and simple. There's a reason it's one of the top three survival priorities. There is a thing called the rule of 3's, that you can survive without food for 3 weeks, without water for 3 days, and without air for 3 minutes. That's how important water is.

Before heading out it is important to fill up or top off your water bottles. I prefer carrying 2-1-liter bottles (I like having a metal nesting cup (like the Toaks Titanium cups) for a Nalgene or a canteen cup for a USGI water bottle, having a metal container will be important if you need to boil water to purify it) for regular camping and I'll add a 3-liter hydration bladder for backpacking and longer day hikes.

Example of a 1-liter Nalgene with nesting Toaks 750ml titanium cup.

In most environments, you need to have the ability to treat water by filtering (like a Katadyn Hiker or Sawyer Mini), purifying using chemical tablets or drops (iodine or chlorine tablets), or boiling from rivers, streams, lakes, and other sources. Getting sick from drinking untreated water is a very real thing and it will take you out of the game quickly, don't do it. In cold/snowy environments you will need a stove, fuel, pot, and lighter to melt snow.

Daily water consumption will vary from person to person. For most people, 2 to 3 liters of water per day is usually sufficient, but in hot weather or at higher altitudes, 6 liters may not be enough. Plan for enough water to accommodate additional requirements due to heat, cold, altitude, exertion, or emergency.

5. Knife

Who likes knives? Probably one of, if not the most, useful tools one can carry! You can use a knife to help with food prep, process firewood and kindling, make other tools with it, cut cordage, make fire, help create shelter, repair gear, make traps to catch food, and more.

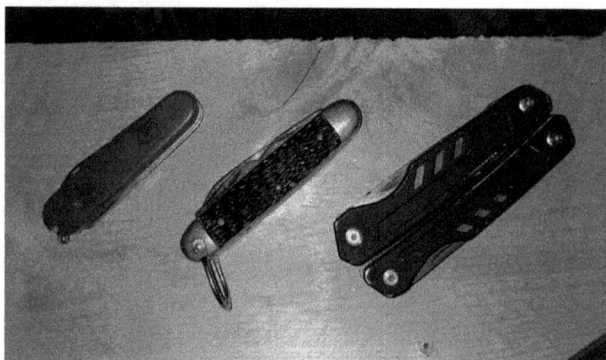

Examples of common pocketknives

So, what kind of knives are there? Classic Swiss Army Knives (SAK), old jack knives, multitools, clip folders, fixed blade knives, etc. There are a lot of knives in the marketplace for outdoor use but what is best for you? Let's cover 3 popular ones.

-Swiss Army Knives: Originally commissioned in the late 1800's, this is a classic Scout knife. Typically, you have a small and large folding blade, screwdriver, can opener and bottle opener, tweezers and a toothpick and a leather awl punch. Easy to carry in a pocket or have a lanyard attached to it. Normally a Scouts first knife is a SAK. Great for basic camp chores like food prep, cutting cordage, and making feather sticks or shavings for starting fires.

-Multitools: Really an upgrade from your classic SAK's in terms of utility. Usually has a single blade, a small saw, multiple screwdrivers, can/bottle openers, awl punch, and built in pliers and wire cutters. Like the SAK it is great for camp chores, cutting cordage, and basic fire prep but also is good with field repairs on gear, the pliers are nice for bending wire or metal for trap and snare making or for fashioning fishhooks from metal. The saw helps with cutting notches for making camp tools and traps.

-Fixed Blade camp knives: There's a reason Jim Bowie carried a big knife. He was a master frontiersman who lived off the land, wrestled a bear, and ventured into the unknown. He needed a knife to handle it all. Fixed blade knife (something with a 3-to-5-inch blade for most use) can do lot. Some look at them as being unwieldy due to the larger size versus a SAK but when used right are more stable and safer. You can baton wood (batoning wood is the act of splitting a larger diameter piece of wood into smaller pieces using your knife or axe), the weight makes creating feather sticks and shavings easier for fire making, making camp tools and furniture becomes easier, and more...

I'm a believer in carrying 2 knives... A small everyday knife like a SAK for in the moment use and a fixed blade for camp chores and harder work. One in the pocket and one on the belt.

The last point about knives is the way we view and teach their use. We need to do a better job of how we view knives. They are tools, nothing more, nothing less. You cut food and rope with them, you create tools and shelter with them. You can make fire to cook and keep warm with them. You can make camp projects like tripods, chairs, cook stations, tables, and more with them.

6. Food

A one-day supply of food is reasonable to keep in one's base kit. Mr. Murphy is always waiting just around the corner to foul up your well-intentioned plans. Weather delays, faulty navigation, washed out trails, injury, and so many other things can cause your 5-hour outing into an unplanned overnighter and longer.

The general rule of thumb is for a day hike to carry a full days' worth of food and on a multi-day trip to pack for as many days as you plan on plus an extra days' worth in case your trek gets set back due to weather, injury, navigation issues, and so on and so forth.

For the type of food, well that gets interesting. My own preference is for easy to eat items that can be carried in my pockets that are high in carbs, protein, and sugars with some salt. This mixture gives you a good mix of energy and electrolytes replacement. So, I like granola bars, beef jerky, hard candies, and dried fruit/nut mixes. For a day hike this works perfect.

Normally, on a multi-day trip my breakfast is simple with a granola bar and a cup of instant coffee, lunch I like to just snack all day on jerky and dried fruits and nuts while on the move, and for dinner I like an easy hot meal like Raman with tuna, chicken, or summer sausage added to it. It's easy, quick, filling, and quick to clean up. This type of menu packs well, requires minimal heating from a small stove, and is lightweight with little waste. I will usually use one water bottle for powdered drink mixes like lemonade.

For the last few years, I've been pre-making ration packs for sustainment kits that are easy to pack. Each ration pack contains approximately 2200 calories including proteins,

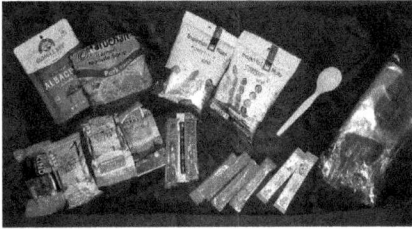

fats, carbs, sugars, and vitamins. 2 of them fit in a pocket on an Alice pack (similar to broken down MRE's for those of you who know) or 3 will fit in a Bergen pocket, once again if you know, you know... If you're going on a day hike toss one in your pack. Going on an overnighter or weekend trip then toss a few more in. The freezer size zip lock bag doubles as a bag for your trash. Bonus is I keep an extra freezer zip lock bag in the ration kit to use for water collection in a survival situation.

7. Extra Clothing

The term "extra clothes" refers to, beyond the clothes on your body, additional layers that would be needed to not just accommodate changing weather conditions, but to survive the long and inactive hours of an unplanned overnighter. So, to begin with you need to ask yourself what extra clothes are needed to survive the night in my emergency shelter in the worst conditions that could realistically be encountered on any given trip?

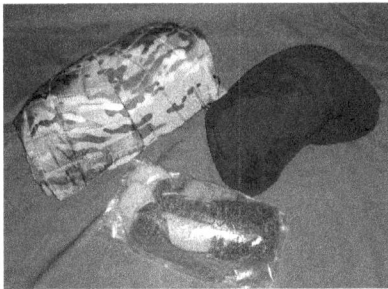

But let's start with some basics for clothing as it is listed in the traditional 10 Essentials... At a minimum in my day pack, I'll carry a spare pair of wool hiking socks, a light insulating layer, and appropriate rain gear. If your socks get damp, you have a spare pair to change into. If you get chilled you have something to put on to warm up with, and if it starts to rain you have something to stay dry in. Seems pretty simple right. But to better understand the notion of "extra clothes" we should probably have a better understanding of the layering system as a whole which we will dive into later in this book.

8. Rain Gear/Shelter

Rain gear... A crucial piece of gear that seemingly gets placed low on many folk's essentials list. We covered extra clothing in a prior post and the importance of staying dry, but yet many still fail to plan for rain. Or if they do, they place their bets on a $1 see through plastic poncho that they sell at Sea World for when Shamu splashes you to stay dry. But it rips the first time they put it on and then look silly standing there soaking wet anyways. Sorry, that just bugs me.

So anyways, proper rain gear is a necessity if you are venturing into the outdoors. For hiking, backpacking, and

camping a well-made rain jacket will keep you dry in the hardest of downpours while also being lightweight and packable when stowed in your pack. Things to look for when buying a good rain jacket are taped seams, adjustable hood, and easily accessible pockets. Pit zips would be a bonus for increased ventilation. Also make sure it's sized large enough to fit over top of an insulating layer.

Another option that is gaining popularity again (and for good reasons) is your classic military ponchos. They are waterproof, small packing, easy to put on/take off, can double was an emergency shelter AND combined with a Woobie can make a lightweight sleeping bag.

Bonus uses for a poncho is being able to set it up during the day to create shade from the sun, make a rain catchment system with it, make an improvised litter, create an improvised pack raft for deep water crossings, makeshift sail on a canoe trip, and more. A military style poncho sounds pretty handy for outdoors use and deserves a spot in the ten essentials list.

It is too versatile a piece of kit to not be included.

9. Fire Starter/Ignition Source

Fire and the ability to make a fire, even in horrible conditions, is one of the big 3 survival priorities (fire, shelter, water). This is why it is included in our 10 essentials list. Having a solid fire kit (and knowing how to use it) in your pack is key to being able to reliably make a fire when it counts most. You and your mates are off exploring and one of you falls into icy water. You have to be able to make a fire NOW. Can You? Do you have the tools to do so? This is why we need a fire kit. With fire you can stay warm, purify water, signal for help, stay safe from predators, cook food, lift morale, see in the dark, and be the envy of your fellow outdoorsmen.

My rule of thumb is to have 3 different ignition sources and 3 different tinder sources.

My ignition sources:
-Lighter wrapped in duct tape
-Lifeboat matches with striker
-Ferro rod with striker

My tinder sources:
-Petroleum coated cotton balls in a metal tin
-Fatwood
-Wetfire fire starters

I also keep a couple small tealight candles for extending a
flame or emergency light.

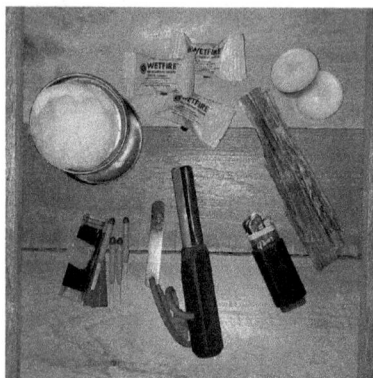

The duct tape on the lighter can extend a flame for several minutes which is useful for making fires in wet conditions. The metal tin can be used for making char cloth or for using as a dry surface for tinder bundles. Fatwood can be scrapped off and put into larger tinder bundles for lighting with a Ferro rod. The resin in the fatwood catches a spark easily and will hold a flame for a few minutes.

The goal with your fire kit should be to make a fire in any situation at any time. Just remember fire prep and practice is just as important, which we will cover later in this book!

The ability to make fire is a skill set that requires constant practice. Fire can provide warmth, safety, signaling, the means to purify water, cook food, and much more.

10. Cordage
Have you ever broken a boot lace on a hiking trip? Ever had a pack strap break? Wish you had a clothesline to dry your gear on? Well cordage can help with those things and more. Keep a 25-to-50-foot hank of it in your day pack.

Cordage comes in all sizes, materials, and strengths.

-Jute twine is common at camp for lashing projects and when fluffed up can be used for fire starting. Usually, a 25 to 50 pound working load at best.

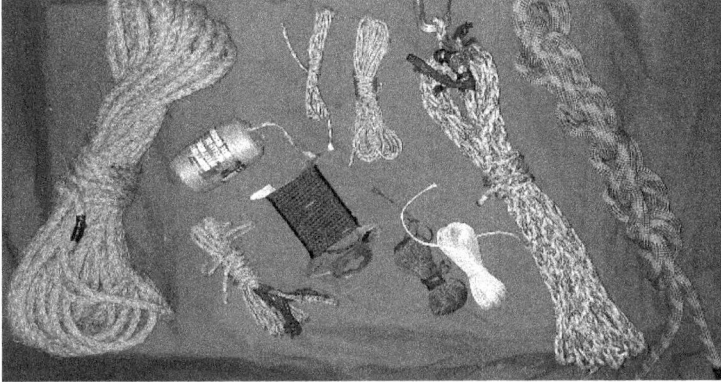

-Paracord (known as 550 cord) is probably the most common type of cordage due to its versatility and strength. Great for boot laces, lashings, tie outs on tarps/tents, ridgelines, clotheslines, lanyards, handle wraps, craft projects and more. Real paracord with the individual inner strands can hold 550 lbs. Hence the name 550 cord.

-Dyneema cord has gained popularity in the ultra-lightweight crowd for its small size and strong weight. It's pricey but can do most of what paracord will do. Some knots do slip through it. The thinner versions as shown have a breaking strength of around 1500 lbs.

-Braided paracord is regular paracord but 3 strands of it are braided together to create a stronger/thicker line for ridgelines or hoists. By braiding 3 lines together the working load increases dramatically.

-Dynamic rope is typically used for climbing. It is designed with a bit of stretch to help absorb the shock of as fall. These ropes go through special certification processes in the US and in Europe to be approved for use and have manufacture dates on them.

-**Sisal rope** is made of natural fibers and has great gripping and knotting strength that will not stretch when wet. Great for tying down loads and gear. Working load is about 250 lbs.

-**Natural cordage** can be made from vines, grasses, inner tree bark, and more. Twisting and braiding pieces together can make strong cordage like native peoples did.

Now for the bad news... Rope work is another "skill" that isn't paid enough attention to either. Knots and lashings are a cornerstone of any outdoorsman's skill set. Time to get creative again with knots and lashings.

Clothing Systems:

Clothing helps you to stay comfortable by creating a thin insulating layer of warm air next to your skin. The enemies of comfort-rain, wind, and cold-work against this protective air layer.

"Comfort" is usually a relative term for anyone who has had to endure harsh conditions. Inclement weather often forces soldiers, outdoorsmen, and here our "Professional Citizen" to endure conditions that can deteriorate far below most people's concept of comfortable. The key to maintaining relative comfort is to stay dry, or when wet, to stay warm and get dry quicky.

But your clothing serves a much greater purpose than mere comfort. In the outdoors, safety is the primary role served by your clothing. When you venture out into the remote outdoors, you sacrifice the option of quickly dashing back to civilization to escape any foul weather. Instead, you have to deal with whatever conditions exist for as long as they last.

For far too many folks, substandard clothing choices have led to hypothermia (*dangerous, uncontrolled drop in body temperature*) that is a leading contributor to death in the outdoors. Your clothing choices should be selected carefully to help ensure your survival through long-term exposure to cold and wet environments.

At the same time, your clothing system should be able to provide protection from overheating on hot days. Breathability and ventilation would be key features to prevent overheating which, in turn, can lead to dehydration, heat exhaustion and heat stroke.

Layering
The layering system consists of 3 main types of layers: Next to skin, insulating, and a weatherproof outer layer.

-Next to skin (underwear) is designed to allow the wicking and evaporation of perspiration keeping your skin dry and therefore warm.

-Insulating layers are designed to trap your body heat keeping you warm. The thicker this layer is, the more body heat it can retain.

-Weatherproof outer layer is your protection from the outside elements. Wind, snow, and rain will rob you of your body heat quickly if you are not protected from them.

Hypothermia is the leading cause of death among outdoor enthusiasts. So, while cotton clothing itself does not kill, it can easily lead to hypothermia — hence why the phrase 'cotton kills' is so often used. When cotton gets wet, it ceases to insulate you because all the air pockets in the fabric fill up with water

How the layering system works
For the above system to work it must be used correctly otherwise you won't receive the full benefit of it. It is called a system for a reason.

BASELAYER MICRO FLEECE LIGHT PUFF JACKET WINDSHIRT BIG PUFFY PARKA

The authors layering system

Cold Weather Layering

It all starts with your base layer. Your base layer is designed to be worn snug against your skin to help in moving perspiration away from your body. As your body heats up it produces moisture and as this moisture sits on your skin it has a cooling effect on it. In the cold weather months, it is imperative to move this moisture away from the body. Examples of good base layer materials would be items like merino wool, polyester, silk, or even for extreme cold environments, a micro or stretch fleece can be worn.

Depending on the temperature and activity levels your insulation layer(s) will differ, but the important aspect of this layer is that it helps create a dead air layer which traps body heat. The thicker the layer the more of your body heat it can trap. But it also aids in the moisture transfer from your base layer away from your body, thus the system aspect of it. Materials here can include microstretch fleece, pile fleeces, light synthetic lofted pieces (think of things like the ever-popular Woobie Hoodie) and more. The idea is the more active the activity then the lighter the layer.

Examples of outer shell layers

This now leads to our outer layer consisting of your weatherproof outer layer. This point of the layering system has been contentious over the years as people learn to understand actual uses of this layer. Everyone knows what Gore-Tex is (waterproof/breathable membrane that lets water vaper escape while preventing water droplets from getting in). But let's face it, if you are out in below freezing conditions, do you really need at GTX shell? Probably not. Do you need to keep the wind out of your layering system so all your hard worked body heat doesn't get blown out, but the perspiration can still escape? Yes. So, a water resistant, breathable piece (Like the Helikon-Tex Windrunner Windshirt or USGI PCU Level 4 or 5 layer) makes more sense! Items like this can make for a great outer layer setup!

The important thing to keep in mind is that the more waterproof an item is the less breathable it will be, and the more active you are being the more breathable you need it to be. The goal is to prevent wind and moisture from stealing away your body heat and chilling your core.

Finally, you need your BPJ (Big Puffy Jacket) for when the stuff really hits the fan. This final layer can be down or synthetic and is great for a warm camp coat or for putting over everything else (size it appropriately to fit over all your layers) when you are taking a rest break so that you don't get chilled, then take it off when you get moving again! These types of jackets are easily compressible for easy storage in your pack and can even be used as an additional sleep layer in an unexpected overnighter (remember the previously mentioned 10 essentials)! I've gotten through numerous unplanned bivouacs by donning a puffy jacket and stuffing my legs into a pack and been able to survive to play another day.

The difficult thing with the cold weather system is when to add or subtract layers. There is a healthy balance with which layers to have on and when. Essentially, when you are on the go, exerting yourself, you will have less layers on. If you are static for long periods of time, then you are adding layers. If it's windy or snowing, you have your softshell or windbreaker on, if it's raining you don your raingear. But essentially, in cold weather climates, the goal is to maintain a constant core body temperature while remaining slightly cool and dry.

Don't forget about your head, hands, and feet! Wool watch caps, balaclavas, and neck gaiters will keep your head and neck warm! Mittens and gloves with enough dexterity for operating weapons and tools are essential, as well as over mitts for days in a camp or on bitterly cold periods on watch!! And your feet, if your feet are in bad shape then you are out of luck! Merino wool or wool blend socks with some sort of wicking liner will help with wicking perspiration away from your feet and will keep your feet warm and dry!

Warm Weather Layering
The layering system for warm weather environments and conditions is sometimes underestimated as to the importance of its impact on the user! Wicking away perspiration, preventing chaffing and blisters, protection from the sun and insects, as well as protecting you from overheating on hot days. Breathability and ventilation are key considerations to prevent excessive sweating, which can dampen your clothing from within and can lead to dehydration and heat exhaustion.

Let's start with the upper body. Synthetic or lightweight merino wool t-shirts are worn as a next to skin layer to wick away perspiration and to help prevent chaffing under chest rigs, plate carriers, and backpacks! This layer is worn close fitting to allow contact with the skin and to allow the movement of moisture away from the body which will also allow the body to maintain its regular core temperature and prevent overheating. Next you layer a looser fitting shirt over top that can provide UV protection from the sun and coverage from insects all the while allowing airflow between it and your next to skin layer. Most folks will use a long sleeve button up or zippered shirt that has a collar to it that can be flipped up for coverage on your neck. Sleeves can be rolled up or left down depending on conditions and user preference!

On your lower body you are following the same rules and wearing lightweight synthetic or merino wool underwear, preferably long boxer brief styles to wick perspiration away and keep your nether regions dry to help prevent chaffing and rashes on the inner thighs and crotch. Durable pants are then worn that are designed with a gusseted crotch for freedom of movement, reinforced seat/knees, pocketed knees for padding. There are many pants on the market today that are made of durable polyester/nylon blends that are incredibly durable, weather resistant, and quick drying! Besides materials and fit, other useful features to look for are large cargo pockets (great for empty magazines and 1st aid items) and extra wide beltloops to accommodate today's larger riggers belts!

Finish up with taking care of your feet with merino wool sock blends to keep your feet dry, comfortable, and blister free along with proper fitting trail shoes/boots that are sized right to accommodate your feet swelling in summer heat and for carrying heavy loads! If your feet are in bad shape due to not being taken care of then you will be out of the game. I use the rule of 3's with my socks! 1 pair on your feet, 1 pair to change into, and the pair that you just changed out of! Keep a small bottle of gold bond for your feet and crotch and you are all set!

Now depending on your AO and conditions, you will have in your pack LW base layers, a lightweight fleece or synthetic jacket, waterproof/breathable foul weather gear, and warm gloves and watch cap... Mountainous areas such as the Rocky Mountain west, Pacific NW, Sierra Nevada's, or New England you can find even summer time temps dropping into the 30's at night with snow and can experience similar conditions in the American Southwest and high deserts! Everyone's needs will be a little different but the above layering systems can be adjusted and tweaked depending on your mission and personal needs!

Equipment:

It is easy to get wrapped up in gear for the Professional Citizen. We waste a lot of time and resources chasing gear when we should be out training in it. So, in this section we are going to cover what we feel (along with many others in our position) are the 3 main pieces of equipment to focus on purchasing when starting off, plus a few extras.

Packs

Packs come in many different designs and sizes, and with that comes many differing needs amongst the end user. The most common question asked is "what pack is best for me?" and that isn't an easy question to answer without first understanding how packs fit into your overall usage and mission profile. You have hydration packs, assault packs, patrol packs, and long range patrol packs. Too many times we get lost in the weeds chasing the ultimate pack instead of focusing on needs. So lets break this down into 3 main pack sizes that make sense for the Professional Citizen.

Patrolling the Eastern Woodlands with a long range patrol pack

-Day Packs (1,000 cu in to 2,000 cu in or 18 liters to 30 liters)
This pack size is great for outings where the bare minimum is needed and for trips where the duration is expected to be 24 hour or less. Maybe it's a quick scout of one's property, a hike with the family, or as an EDC bag for your vehicle. Its volume is enough for an insulating/rain layer, spare socks, snacks, water, first aid, and your remaining 10 essentials as mentioned above. These packs usually have a thin polyurethane frame sheet or pad and no hipbelt and are best with loads weighing under 20 lbs.

-Patrol/Assault Packs (2,200 cu in to 3,500 cu in or 35 liters to 50 liters)
The Patrol or Assault Pack is probably going to be the most used pack for the Professional Citizen. Perfect in size for 2-to-3-day trips that require multiple days' worth of rations, fuel, sleep and shelter kits, foul weather gear, mission

essentials such as observation gear, and possibly more. A lot of folks will consider this family of packs when putting together GHB's (Get Home Bags) or BOB's (Bugout Bags). Many of these packs will have a thin frame sheet with some sort of aluminum internal frame and adjustable hip belt to better help with weight transfer to one's hips. This is due to the design being able to carry larger/heavier loads than the above-mentioned daypack. To be able to carry 30 to 50 lbs. would not be uncommon due to the nature of the use of the pack.

-Long Range Pack (3,600 cu in on up or 60 liters plus)
If you need a pack to support trips in excess of 5 days (or
shorter cold weather trips), then these larger packs are just
what you will need. Lets use a 5 day outing as an example...
That's 5 days worth of rations (6 if you're into contingency
planning), medical, communication and observation gear,
clothing/layers, spare ammo, E&E gear, and possibly more
depending on the scope of that outing. Cold weather trips
require additional layers and bulkier sleep kit not to
mention the need for additional calories for sustainment.
All of this comes at a weight and size cost hence the need
for larger volume packs. It is not unheard of to need an 80-
to-110-liter pack for a 5 to 10 day winter unsupported
excursion. These packs will have large external pockets,
expandable lids, more robust harness systems, and heavier
duty frames to help take the brunt of the excess weight to
be carried.

In a perfect world you would have all three, but it isn't a
perfect world so you should focus on your most likely use
and get the best that you can. Generally, most of our needs
will fall in the first two options so this simplifies things a
bit. For the budget minded you can never go wrong with a
Medium Alice pack. 40 liters in size with 3 large external
pockets and the ability to add a couple sustainment pockets
if needed. For under $100 you have a very serviceable pack
that has a proven track record. If the funds allow it, I would
recommend options like the Mystery Ranch 3 Day Assault
Pack or Karrimor Predator Patrol 45 Pack.

Sleeping bags and Sleep Systems
Sleep is important no matter the time of year but becomes
even more so in colder weather. Your sleep system really is
made up of 3 main things: a sleeping pad, your bag, and
your cover.

Sleeping pad: This is not only to provide cushioning but
also to prevent heat conduction loss with the ground. The
colder ground will suck the heat from your body, regardless

of the rating of your bag, if you don't have adequate insulation under you. You will see the term "R-Value" used when describing a sleeping pad's thermal ability to resist cold traveling through it. The higher the R-Value then the warmer the pad will be, just as in home insulation. That lightweight inflatable pad used during the warmer months may not be enough for the winter temperatures. The colder the temps then the more you need under you.

My setup for the colder months is to double up on pads by using a closed cell foam pad like the old GI foam pads or ThermaRest Z-Fold pads with an inflatable on top. This gives me a couple extra inches of insulation under me without adding too much bulk to my overall sleep system.

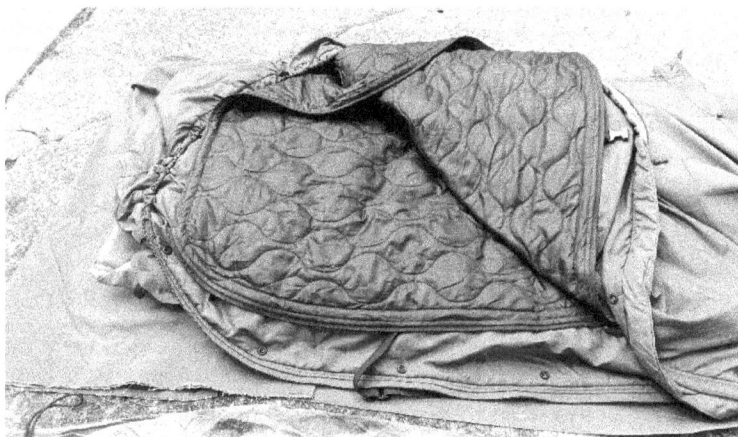

GI Patrol bag lined with a poncho liner.

Sleeping Bag: Your insulation comes from your bag(s). Down, synthetic, wool, etc. are all popular choices with each having their own pros and cons.

Down- Pros: High warmth to weight/bulk ratio. Highly compressible, lightweight, breathable, long lasting when cared for, can be refilled. Cons: Expensive, tough to dry in the field if it gets wet (even from your own perspiration), will not retain warmth when wet, durability.

Synthetics- Pros: Cost effective, durable, maintains some warmth when wet, easier to dry (still sucks to do in the field), widely available. Cons: Can be bulky to maintain same warmth ratio as down counterpart, heavier (but has improved through the years), once the synthetic fibers start breaking down it needs to be replaced (durability), breathability.

Wool- I have wool added here for folks who enjoy using traditional cowboy/sleep rolls. Pros: Maintains warmth even when wet, breathable, durable, flame resistant, widely available, can be inexpensive with many military surplus options. Cons: Heavy, bulky, can be expensive, difficult to find good wool blankets.

Loft equals warmth in a sleeping bag. The more loft (thickness) there is then the more body heat it is able to efficiently trap, therefore keeping you warmer. Sleeping bags do not generate warmth, they trap your body heat as it radiates away from you. A mummy style bag that follows the shape of your body will also do a better job of this versus a large rectangular bag (less dead air space in the bag to heat up). This all plays a role in choosing the best bag for your needs.

Keep in mind that temperature ratings on a bag are a guideline, not an absolute. Some folks sleep colder/warmer than others. Bag ratings take into consideration a person of average build using a 1/4" foam pad and sleeping in a tent. But you need to consider your body type, metabolism, caloric intact/expenditure during the day, and more.

Cover: This is your weather protection layer or the outermost layer of your sleep system. For the type of field outing we usually discuss, most of us aren't utilizing tents except in special circumstances, so some sort of bivy sack is required to protect our sleep system from wind, rain, and snow. Once your bag gets wet, you're done and out of the game. There are a lot of options here with the most popular being the USGI Bivy Bag. Plenty roomy inside for winter weight bags and a pad. Yes, you read that right, your pad too. Most of us move around in our sleep and find ourselves off our sleep pads at some point in the night and get woken up by that cold spot that developed due to being off our pad. Putting your pad inside your bivy keeps your system all together throughout the course of the night keeping you warmer and more rested. So, buy a bivy accordingly to fit your sleep system. Bivys are also great in the warmer months with just a Woobie (poncho liner) inside making an effective summer sleep setup. But beware, while bivys keep outside moisture out and are somewhat breathable, you can get condensation build up inside and this can wet your bag out on multi day outings, making it important to dry or air out your sleep system when safe to do so.

Footwear
This is an item rarely talked about and I think that it is due to the amount of variables in selecting the right footwear. Folks ask all the time, "what is the best boot?", well that is hard to answer because everyone's feet are different.

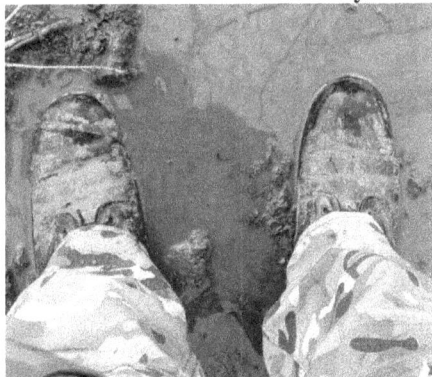

Different foot widths, volumes, high instep/low instep, long arch/short arch, and so on and so forth. So you can see why it is difficult to discuss footwear. But there are some certain truths when looking at boots that you should know about.

-Fit
When going to try on boots make sure you do so with the socks you plan on wearing along with any orthotic devices,

insoles or other inserts you plan on wearing in them. Plan on trying them on towards the end of your day when your feet are swollen from being on them all day long. If possible, take a loaded pack with you to wear while trying boots on. See how they feel under a load. Does the arch support still hit in the right spot? Are your toes getting pinched when the boot is flexed? Do your heals stay firmly seated in the heal cup? It is better to find these things out in the store than while out on the trail or in the field.

-Socks
Your sock choice can make or break an outing. Socks are designed to cushion and insulate the feet and reduce friction between your foot and the boot itself. They must also be able to wick perspiration away from your feet and allow them to breathe. Materials such as merino wool or polyester blends work well at moisture movement and dry fairly quickly as opposed to cotton which gets saturated from sweat and stick to your feet leading to blisters. I am a firm believer in the 3 sock rule meaning 1 pair is on my feet, 1 pair is dry, and the 3rd pair are the ones that I just took off and are drying. This 3 pair rotation is a minimum for socks in the field to help maintain dry, healthy feet. Once your feet go then so do you. Don't let your feet get ruined.

-Boot Care
With proper care, good boots can last for several years. Always use the manufacturer's recommended boot conditioner and make sure you keep your boots clean and dry when not in use.

-Soles
I highly recommend finding boots with stiffer soles that, while offering natural flex while walking do not have any counter bend to them when stepping on rocks or tree roots under a heavy load. This can damage the arch in your feet and can take a long time to recover from. A simple test is when trying on boots to throw a golf ball size rock on the

ground and step on it. If you can feel the rock underfoot then the midsole of the boot may not be stiff enough.

-Ankle Support
With the advent of ultralight backpacking came lightweight trail shoes. Makes sense, you're carrying a lighter pack (sub 20 lbs.) so your support needs go down as well. But, in an environment with unmaintained trail and heavier loads, support becomes important. Remember, if your feet fail then so do you. Don't be a liability to yourself and your family or team by trying to skimp on foot protection because the ultralight crowd told you to do so. They're playing a different game with different consequences. When carrying loads on uneven terrain, the likelihood for ankle injuries goes way up, especially when you are exhausted, hungry, wet, and under possible distress.

-Waterproof/Breathable vs Non-Waterproof
Your feet sweat, A LOT... They will sweat more inside a boot with a waterproof/breathable membrane than a boot without a waterproof membrane. Further dampening your socks and necessitating increased sock changes. Some folks this isn't a concern but for others it is. A non-waterproof lined boot can be made highly water resistant by using the appropriate waterproofing agent but still has the natural breathability of that boot. This is a personal decision that you must make the call on and only your own experiences can steer you in one direction or another.

-Gaiters
Out in the field, water, snow, and debris can get into your boots by going in over the tops of them getting your feet wet and uncomfortable leading to blisters. Boot gaiters are designed to seal the area where your boots meet your pant legs. Gaiters also help prevent your pant legs from getting wet when walking in deep snow or wet underbrush. Having a snug fit around the boot is essential to prevent snow and debris from entering under the gaiter as well.

Gaiters are great for preventing snow from entering the tops of your boots

In the end, boots are a very individual purchase and is something not to skimp on. Good boots are your mode of transportation in the field and taking the time to find the right pair is important.

Stoves
Cooking in the field can take place in many forms and can be determined by your food planning, how permissive your environment is, time constraints, and time of year/weather considerations.

Many times, your food choices will dictate the type of stove you need. You have liquid fuel stoves, canister fuel options, alcohol, and even small folding stoves that use fuel cubes or biomass (twig stoves). So let's break them down.

Heating water for coffee

-Liquid Fuel Stoves
These are really your workhorse stoves. Great for actually
cooking and for boiling large amounts of water and melting
snow/ice in winter conditions. These are ideal for higher
altitudes and cold temperatures due to being able to
control the pressure of the fuel. Many liquid fuel stoves
work on not only white gas but also kerosene, gasoline,
diesel, and even jet fuel. Extremely versatile but can be
loud, heavy, and not as compact as many other options.

-Canister Fuel Stoves
Renowned for their compact size and light weight. JetBoil
and the MSR Pocket Rocket are probably some of the most
popular options amongst outdoor enthusiasts. Great for
boiling water for rehydration dehydrated/freeze dried meal
options and for quick cups of coffee. Most stoves in this
category can nest inside of numerous different cook kits
such as the ever-popular Stanley Cook Kit or inside of

European surplus cook sets. The fuel used is a propane/isobutane mix that burns fast, hot, and is very light weight. This category of stoves is perfect for the minimalist who is only heating water for warm beverages and food rehydration. Unfortunately, the fuel canisters, while they can be refilled, end up being disposable which means carrying waste with you on the trail. They can also have a bit of a noise signature.

-Esbit Style Stoves
The old school Esbit style survival stoves are incredibly small, lightweight, and utilize solid fuel cubes or twigs for fuel. They are quiet, compact, and easy to use for warm drinks and soups. The old GI Canteen Cup stoves would be in this category as well as other folding twig stove style setups. These have gained popularity recently as folks are trying to slim down kits for size, weight, and simplicity.

Canister fuel stove with Toaks Titanium cup

Cook Kits

Cook kits come in all different shapes and sizes. I like having self-contained setups that contain everything I may need. From old canteen setups with canteen cup, lid, and cooker to a minimalist Nalgene nesting cup and Esbit stove or the ever-popular German style mess kits that are extremely versatile.

Examples of cook kits.

My suggestion is to choose a cook setup based on your needs. If you are doing actual field cooking then a titanium cup isn't going to be a wise choice, you may want something more robust to tackle that type of use. This is where something like the older surplus cook kits (like one of the European style kits) would be preferable. Especially if using over open fires and on coals. It is designed to take the abuse.

44

If all you are doing is heating water for coffee and rehydrating backpacking meals, then a classic canteen cup setup is quick and simple or the newer nesting style cups (such as the Toaks titanium or GSI stainless steel ones) work great and take up very little room. If you want something a bit more substantial the Stanley Cook Kits are a popular item as well.

If you could smell and taste a picture

Only you can really decide which will work best. I recommend having a couple of different cook setups so that you can pick and choose based upon your needs. Don't be afraid to experiment with different setups. If all you have is a GI Canteen Cup setup, don't worry, It is a tried and true kit that has worked around the world. Get out there in the field and use them!

Fire:

If you haven't worked with fire a lot, then it is hard to express the emotional connection that it gives so many of us. But the benefits are easy to understand. Fire can give us light and warmth. It can boil our water and cook our food. It can scare away things that go bump in the night. To say the least, fire, and the ability to make it, is one of our most important tools in the wilderness.

Sometimes a campfire just sets the mood

Fire needs three basic ingredients to be brought to life: oxygen, heat, and fuel. Oxygen is everywhere around us in the air that we breathe, so that leaves us responsible for the other two ingredients. Heat comes from our ignition source (this can be a match, lighter, ferro rod, or even from friction), and the fuel comes from our surroundings. In the field our fuel needs to be dry material and in order to properly build our fire our fuel sources need to be collected and staged. There are 3 sources of fuel:

-Tinder
Your tinder is going to be your finest of combustible materials. Examples would be wood shavings, leaves, pine needles, dried grasses, essentially anything light and airy. You want to form your tinder into a bundle resembling something like a birds nest. This shape allows for easy lighting or catching of sparks. In wet conditions you may need to add an accelerant to aid in the lighting of your tinder bundle, think of things like fatwood shaving, petroleum cotton balls, or any number of other premade fire starters.

Tinder bundle made from a feather stick and red cedar bark

-Kindling
Your second layer for your fire is your kindling. This is the material that will feed the flames of the tinder and give your fire enough heat to burn larger pieces of wood. Kindling should be dry enough that when you go to break it there is an audible snap to it. Size wise, I like to think of kindling as being anywhere from the thickness of a pencil to the thickness of your thumb. There will be no such thing as having too much kindling.

kindling examples from pencil size to thumb size

-Firewood

Your firewood is meant to sustain your fire. Size wise it should be anything sized up from your kindling up to forearm sized wood. Gather as much as you can to keep your fire going. If you think you have enough wood, then you need to go collect even more. You will never have enough.

The first fuel that our fire will eat is the tinder, which burns fast and will generate the heat needed to ignite our kindling which in turn creates embers which allows for the burning of our firewood. Once our firewood catches it can burn indefinitely as long as we keep feeding it more firewood.

But none of this happens without having an ignition source, so having a proper fire making kit is important in being prepared for making your fire. I like the rule of 3's when it comes to ignition sources. My three favorites are waterproof matches, a lighter, and a ferro rod. To go along with this, I like to keep three types of Firestarter options in my kit as well. My go-to options are fatwood, PJ cotton balls, and fire plugs.

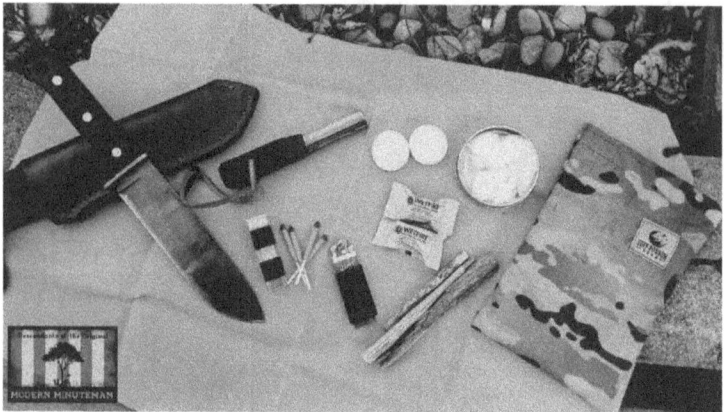

Example of a well-prepared fire kit

Lastly, fire prep is key in ensuring you can get a fire lit and maintained. Once again, the rule of 3's applies here when collecting your materials. You want to make sure that you collect enough tinder and kindling for making 3 fires (in case the first attempt or two fail and for one the following morning) and enough firewood to last all night long.

So, lets build a fire...

Probably the easiest fire lay to make is the classic twig fire. It's simple, quick, and effective. It all starts with having a clear and level area for making your fire and all your fire preparation within easy reach.

Proper fire material prep

Layout your tinder bundle and light it. Once you have
flames grab a loose handful of kindling (smaller the better)
and loosely place them directly on top of your tinder
bundle being careful not to smother it. You are almost
creating a loose tepee over your tinder. It may take 2 or 3
handfuls of kindling to get enough heat to start burning
your firewood. Add a piece or two at a time until you have a
sustainable fire. Add firewood as needed.

A fire slowly comes to life

There are many other types of fire styles and once you catch the fire making bug you will spend hours on end practicing your newfound skill!

Food:

Food is not often talked about in field manuals beyond MRE's and Mountain House meals, but it is vitally important to your existence in the field.

Sure, we all wish we had a fully stocked field kitchen (you do have one, right?) But that isn't always going to be the case. You are going to have to have a means of sustaining yourself while out on patrol or manning an OP/LP. MRE's and Backpacking meals are nice but can get expensive and may not always suit one's tastes or preferences. So, after years and years (decades really) of planning rations for trips I have realized that the best ration kits are the ones you put together yourself.

Calories are king when it comes to working in the field. The goal is to be able to replenish the calories you burn during the day. The harder you work, the more weight you carry, the time of year, and so on and so forth will determine your caloric needs. The average male who is sedentary can get away with 1,500 to 2,000 calories per day. An active male may require 2,000 to 2,500 calories per day depending on your own personal needs. This is a general guideline as we all will vary. I know that I tend to be on the light side of caloric intake when I am out in the field, it's just the way my body works. I have friends who go lighter on calories and others who go heavier. You must know your own body and your own needs when food planning.

Now this is generally how I eat throughout the day.

Breakfast: I'm not much of a breakfast person at home so the same goes for on the trail. I usually have a cup of instant coffee or two and a granola bar of some kind. This is enough to get me going in the mornings.

Lunch: Snacking commences an hour or so out of camp. I will just lightly snack on a mix of jerky/Slim Jims, nuts, granola bars, and dried fruit/nut mixes as the day goes on. This doesn't require any long stops and can be easily eaten while on the move. I will also keep one water bottle filled with some kind of electrolyte mix to drink as well.

Dinner: This is my one and only "meal" for the day. It is usually late and requires minimal cooking or none at all. I fancy Ramen with tuna or chicken mixed in with it or rice meals with beans and tuna/chicken packets. These are easy to make and can be heated in a canteen cup. Sometimes a specific outing may require no cooking where I will then substitute out the Ramen for precooked rice/tuna bowls like the Starfish Smart Bowls.

Each days worth of rations are kept in a large zip lock bag and includes a disposable spoon and an extra zip lock bag that can be used for garbage or for water collection. You could also add daily vitamins, individual wet wipes, or even gum or candy to these kits for a bit of extra convenience in the field.

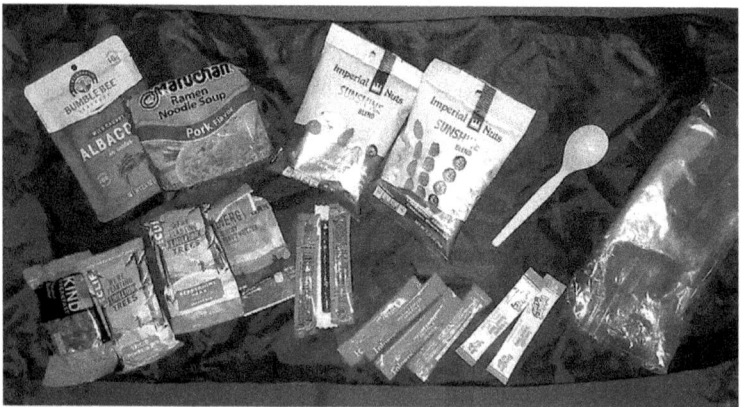

Example of a 24-hour ration pack

I keep these stored in 5-gallon sealed buckets at home and when heading out can grab 1, 2, or more of the ration kits inside depending on the length of the trip. I have found that 2 ration kits will fit in an ALICE pack pocket and 3 will fit in a Bergen side pocket. These also work great in prepackaged sustainment kits for delivery to the guys in the field. Quick, simple, and effective!

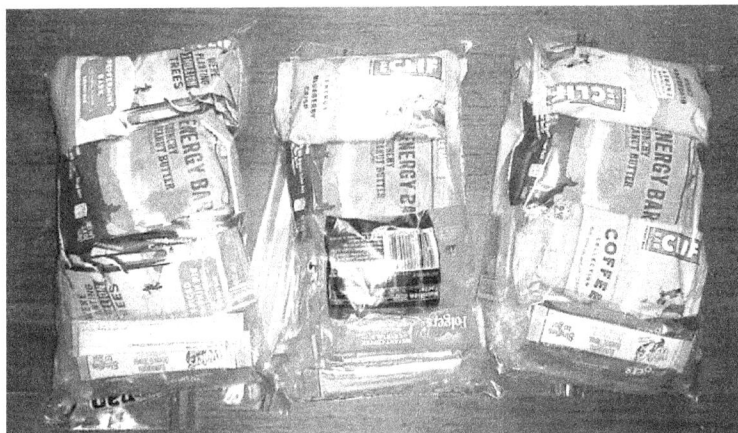

Example of 3 days' worth of rations

At the end of the day, it's up to you how you want your chow in the field, and I've learned through my own experiences that simple is better for me. The above is a tried-and-true system that has never let me down, whether it be on the side of a mountain, on a backpacking trip, or out on a property patrol or hunkered down in an OP, it just works!

Water:

You need water, without it you will die. Plain and simple. There's a reason it's one of the top three survival priorities. There is a thing called the rule of 3's, that you can survive without food for 3 weeks, without water for 3 days, and without air for 3 minutes. That's how important water is.

In most environments, you need to have the ability to treat water by filtering (like a Katadyn Hiker or Sawyer Mini), purifying using chemical tablets or drops (iodine or chlorine tablets), or boiling from rivers, streams, lakes, and other sources.

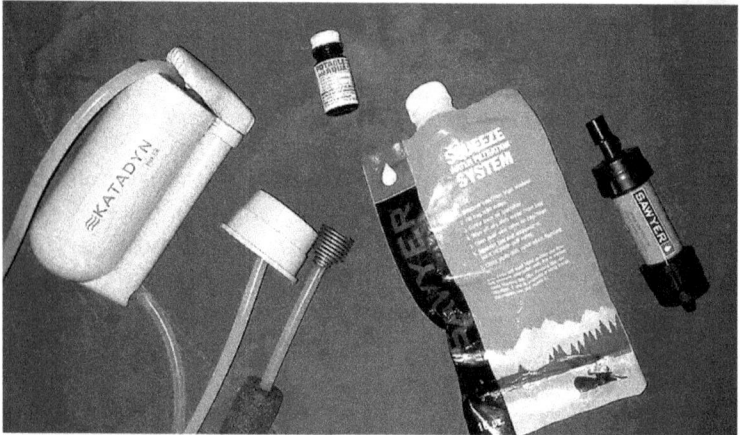

Examples of popular water treatment options

Getting sick from drinking untreated water is a very real thing and it will take you out of the game quickly, don't do it, your buddies will hate you for it. In cold/snowy environments you will need a stove, fuel, pot, and lighter to melt snow.

In a true survival situation you need to know how to procure water from your environment via non-traditional methods. You may need to dig a hole to reach the water table, look for vines or trees to tap into, making a field expedient gravity filter utilizing sand, grass, stones, and charcoal, or even creating a solar still. These are important skills to learn and practice.

Daily water consumption will vary from person to person. For most people, 2 to 3 liters of water per day is usually sufficient, but in hot weather or at higher altitudes, 6 liters may not be enough. Plan for enough water to accommodate additional requirements due to heat, cold, altitude, exertion, or emergency.

Examples of 1-liter bottles with accompanying nesting cups

Travel:

What may seem to be the most basic of activities is a skill we do every day: walking. But just because you have the ability to talk doesn't make you a public speaker, the ability to walk doesn't make you a backcountry traveler. To be able to efficiently navigate the backcountry you must consider the terrain, weather, pack loads, physical conditioning, and the overall safety of your surroundings.

Established trails are, obviously, the simplest way to explore the backcountry. They usually lead to well-known spots, campsites, water sources, and for better or worse, other people too.

Doing your homework ahead of time can help alleviate the stress of route finding and navigation. Getting detailed maps of your area and premarking areas of interest, terrain features like water, river crossings, cliffs, roads, population centers, and more. It is important to identify these things ahead of time so that you know what you are getting into and can have the appropriate plans and equipment in place to safely navigate it. Then you need to get out into the area and start exploring. In many areas you may need to navigate by map/compass, in others you may be able to rely on terrain features for route finding. Intuition and luck can also play a role but there is no substitute for firsthand experience.

Stay alert to hazards. Study waterways, roads, snowfields, cliffs, and other exposed areas. Continually evaluate possible hazards and look for continuous routes. If something looks questionable, search for alternatives and make your decisions on alternatives as early as possible.

This is also a time to look ahead to where you want to be by the end of the day. Consider where you want to be by the time darkness falls and consider if it is safe to travel at night. Also stay on the lookout for emergency overnight positions, water supplies, and egress routes if anything were to go against your well-intentioned plans.

Pace
One of the hardest things for beginners is finding the right pace. Folks often make the mistake of either going too fast or too slow.

The most common mistake is going too fast. Partly out of concern for the distance ahead or they feel the need to outperform those with them. The correct pace is the one that you can sustain for hours on end, and the only way to find this out is to go out and put those miles under your belt. Your pace will change throughout the day depending on the terrain, weather, the weight of your load, and as fatigue sets in. Catching your "second wind" is a real thing as adrenaline sets in but remember that there is no such thing as a "third wind".

Rests
The ability to take brief rest stops throughout the day is a necessary thing. This can allow you to catch your breath, have a quick snack, and get something to drink. Some rest breaks can be coordinated with areas for refilling water, as terrain starts to get more difficult, or as the noon day sun is starting to beat down. Sometimes if on coordinated movements, a brief rest can be initiated while group leaders pull aside to get oriented to the map, to figure out the next course of action, or to work out group issues.

Keeping breaks short prevents one from cooling down too much (in cooler weather) and from getting stiff and cramping. I have found that frequent 5-minute breaks (keeping gear on) every hour helps with staying fresh and alert and then a pack off break every 2 hours for reenergizing helps with morale and sore bodies.

Special Considerations

As we journey into the backcountry, there will be many things that you encounter that will require extra thinking and considerations as you plan on how you will navigate them. Things like trail finding, bushwacking, talus and scree slopes, snow fields, stream crossings, and many more. Many of these will become hazards that you will want to weigh certain pros and cons as to how you decide how to proceed. There will be many risk vs reward decisions to make and in many cases experience will help you deal with these things.

Traveling in the wilderness can be a complicated endeavor because of all the variables of the time of year, weather, terrain, water, snow and more. Combine this with the other information laid out in this chapter concerning food, water, equipment, and clothing and you should feel good about starting off on this journey.

Hygiene:

Good hygiene is important in the field. Small things become bigger things and can jeopardize your health. Left unchecked you can go downhill fast. Let's look at a few important items to be wary of.

Your Responsibility: Everyone is responsible for their own hygiene, plain and simple. It isn't your mother's job to remind you to wash your hands, brush your teeth, and to comb your hair. It's on you. Yes, you are in the field, many times with your buddies, but you still need to stay clean.

Personal hygiene: Folks should have some sort of a personal hygiene or toiletry kit in their gear. This can be extensive for field living in a base camp or forward camp and can be tuned down for quick outings and field exercises. Your kit will include, but is not limited to the following items:

*Absorbent body powder.
*Alcohol-based hand sanitizer.
*Antiperspirant/deodorant.
*Comb.
*Dental floss.
*Insect repellent.
*Eye drops.
*Feminine hygiene products.
*Foot powder.
*Hairbrush.
*Lip balm.
*Prescription medications (for example, birth control, blood pressure, and so forth).
*Sanitizing wipes.
*Shampoo.
*Shaving kit.
*Soap.
*Sunscreen lotion.
*Toilet tissue.

*Toothbrush.
*Toothpaste.
*Towels.
*Washcloths.

SKIN CARE
The skin is the largest organ of the human body and protects the body from disease-causing bacteria and viruses. The skin also provides protection from the direct rays of the sun, insulates the body from cold, and helps to regulate the temperature of the body in hot environments.

As the body's first line of defense it is essential that folks protect their skin by keeping it as clean as possible. Showering regularly helps to reduce bacteria that are resident on the skin and can help to prevent infection from scrapes, cuts, punctures, and cracked skin. Folks can protect their skin by:
*Showering or bathing regularly to keep the skin clean.
*Using absorbent body powder to control moisture buildup. Pay particular attention to areas where wetness is a problem (such as underarms, between the thighs and buttocks, feet, and, for females, under the breasts).
*Wearing proper clothing and layers
*Wear moisture wicking undergarments designed to pull moisture away from the skin.
*Changing into clean dry socks and applying antifungal foot powder to protect the feet from prolonged periods of dampness.
*Applying insect repellent when needed.
*Applying sunscreen to exposed skin.

SHOWERING
Under ideal conditions folks should want to shower daily, or at a minimum once every week to maintain good personal hygiene. Frequent showering prevents skin infections and helps to prevent potential parasite infestations. When showers are not available, washing daily with a washcloth and soap and water is advised.

Particular attention should be given to sweaty areas or places that become wet armpits, feet, genitals, between thighs and buttocks, and under breasts.

In situations where shower facilities are not available, a full canteen of water should be adequate for one person and a five-gallon water container for small groups. While it would be nice to have hot water for washing in the field it may not always be possible. The hot water wash is especially refreshing after a long day on the field.

Folks should avoid using perfume, cologne, or scented soaps, which can attract insects and alert others to your location. Scents travel a long way in the backcountry so we need to be especially wary of this. However, unscented lotion can be used to keep the skin from drying, cracking, and becoming infected.

HAND WASHING AND SANITIZING

One of the most effective practices that folks can perform to protect themselves and others from the spread of disease is to thoroughly wash or sanitize their hands frequently. Regular washing or sanitizing of the hands denies disease-causing bacteria and viruses from gaining easy entry into the body. Folks who fail to wash their hands frequently increase the risk of spreading germs picked up from other sources and possibly infecting themselves when touching their eyes, nose, or mouth. One of the most common ways folks catch a cold is by rubbing their nose or their eyes with an unwashed hand which has been contaminated with a cold-causing virus.

Germs can be spread directly to others or onto surfaces that others might touch which may cause other folks around you to become sick. The important thing to remember is that, in addition to colds, serious diseases like infectious diarrhea and meningitis can easily be prevented when folks make a habit of frequently washing their hands.

When to wash and or sanitize the hands (at a minimum):
*Before eating or snacking.
*After eating or snacking.
*Before handling or preparing food.
*After using the latrine.
*After handling anything that could potentially transfer germs.
*Frequently during the day to keep your hands free of germs.
*After coming into contact with any local flora or fauna.
*After physical contact with local nationals.

I'm sure we have all seen the survival show Alone, right. Well one of the biggest mistakes folks make on the show that takes them out is poor hygiene practice, especially simple things like just cleaning their hands. It is that basic and yet that important to do.

Ways to clean or sanitize the hands is through the use of:
*Soap and potable water.
*Alcohol-based hand sanitizing solutions when soap and water are not available.
*Commercial cleansing wipes if available.

ORAL HYGIENE
The issue of oral hygiene is a readiness issue. Folks who fail to maintain a vigorous oral hygiene regimen can quickly become a risk to your group. When neglected bacteria in the mouth use starches and sugar to produce acids that can quickly result in gingivitis and tooth decay. Not brushing for just a few days can cause inflammation of the gums and result in irritated and bleeding gums. If gum disease already exists, it can quickly become worse. To prevent tooth decay and gum disease The Professional Citizen must maintain good oral hygiene practices at all times by:

*Flossing their teeth.
*Brushing their teeth.

Flossing is important because it removes food particles from between the teeth and under the gums where brushing cannot reach. Folks should floss at least once per day.

Folks should brush at least twice a day, especially before sleeping. Brushing should include the use of fluoride toothpaste to brush all the surfaces of the teeth using a circular motion. Folks should not rinse, eat, or drink anything for at least 30 minutes after brushing to allow the fluoride to stay on the teeth longer and protect them better. If toothpaste is not available, folks should brush their teeth anyway. Brushing should include the tongue and the roof of the mouth. Folks can also enhance their oral hygiene by chewing the gum contained in the accessory packet of every field ration. The gum is made with a sweetener that helps control the buildup of oral bacteria and reduces tooth decay when used regularly.

Folks should brush regularly even when running water is not available. While in the field this can be accomplished by keeping a small toothbrush in a ventilated toothbrush cover or case and kept in a convenient pocket.

Note. After brushing, rinse the toothbrush by pouring a small amount of water over the bristles.

If a toothbrush is not available, folks can rinse their mouths with water after eating then wrap a piece of cloth around a finger and wipe the surfaces of the teeth and gums.

Field Latrines
I'll dig into the old backpacking, leave no trace philosophy on this. Having to pee and poop in the field is a necessary chore. It is what it is, but there are some best practices to consider when doing so.

In a base camp or group camp a designated area needs to be established for a latrine. Ideally, your latrine area should be a minimum of 150 to 200 feet away from any water source, trail, and camp area (making sure to have it downwind of your camp).

Trench toilets and Cat Holes will be the two most common types of field latrine setups. Key items are going to be:

*An E-tool or trowel
*Toilet paper
*Wet wipes
*Hand sanitizer

The infamous Cat Hole is a field expedient way of making a quick latrine spot for human waste. Using your E-tool or a small trowel, you will need to dig a hole approximately 12" wide and 6" to 12" deep. Keep the disturbed soil to the side

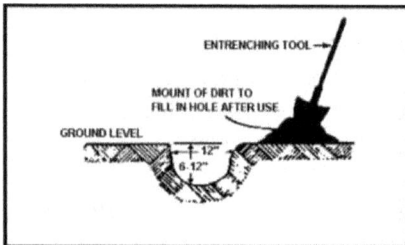

for filling back in the hole once finished. A best practice is to keep your toilet paper, wet wipes, and hand sanitizer in a zip lock back to keep it separate from everything else.

The trench toilet is well suited for camps where multiple people will be utilizing it. Find a suitable area as we stated above and dig a shallow trench approximately 3' wide and

6" to 12" deep. Just as with the Cat Hole keep the soil piled next to it to cover your waste when finished. Best practice is to erect something to lean against (see image) while in use and keep it in an area that offers some degree of privacy.

Notes:

CHAPTER 2
Emergency Prevention and Response

Why do we need to discuss Emergency Prevention and response? Well it is a cornerstone of what we do. Planning, risk assessments, leadership, safety, rescue, and so much more. These are not often talked about amongst the Professional Citizen, but are extremely important to the success of the greater cause.

Leadership:

"Leadership is the capacity to influence others through inspiration motivated by passion, generated by vision, produced by a conviction, ignited by a purpose."

— Myles Munroe

Every group, regardless of its size or nature, needs not only a leader, but good leadership. It's one thing to go out on a field exercise with a group of well-seasoned friends but it is a whole other thing to bring people together that you don't know to a place none of you are familiar with. You and your friends achieve this naturally without even realizing that you are doing so, but the other group requires a more formal, even structured organization. Either way, the role of leadership is the same, it provides a way of putting a field exercise or trip together to make it safe and successful for everyone involved.

Usually, small informal groups do not need to select a leader. Everyone shares in the responsibility for the work, the organization, the logistics, the planning, and the carrying out of the overall goal. Everyone knows what the others are doing and not much discussion is needed for getting things done. Everyone tends to be a "been there, done that" type of person so everything just naturally happens.

67

Larger groups, however, do not lead themselves as easily and require some formal leadership. Someone who can organize duties, delegate workloads, to implement team building, and so much more.

Leadership types can be broken down into a few different categories...

-Peer Leader
This is most common amongst small groups of friends who decide to head to the woods for a weekend. Members of this group tend to informally decide who does what concerning food, equipment, transportation, activities, and more. In todays world a lot of this is done via text messages throughout the week leading up to the planned weekend. Even in this group one person will rise above the others as the "leader" through general consensus.

Example of a peer led group

-Group Champion
This is the guy (or gal) that "champions" the outing, the one with the original idea and invites everyone else. While

never officially named the leader, they are the one that everyone recognizes as the one in charge.

-Most Experienced
This tends to be common in the climbing world, the person with the most experience is bestowed the leader by everyone else in the group.

-Course or School Leader
This would be the case if you were to sign up for a guided climbing trip or partake in any outdoor skills course. Leaders here are usually accredited through a formal organization or school to ensure that they have a certain level of experience, knowledge, and competence.

Leadership isn't just there to say I'm in charge. It isn't an ego trip or a chest thumping position. It can be one of the most rewarding and fulfilling experiences as well as one of the scariest.

Leading a group in the Adirondack Mountains

Leadership Roles
Not only is the leader in charge of the group, but they are also the teacher, trip planner, subject expert, coach,

mentor, guardian angel, decision maker, arbitrator, and dish washer.

-Teacher
When you have less experienced folks along with you, it is important to seize on teachable moments to make sure that everyone is on the same page with what is happening. Usually it is nothing more than advice and demonstration. When safety is concerned, you may end up taking the time to have a classroom session on how certain things are done. I'm a fan of the EDGE method when it comes to this, **E**xplain, **D**emonstrate, **G**uide, and **E**nable. This is also referred to as Servant Leadership, where the well-seasoned outdoorsman is willingly sharing and teaching their hard-earned knowledge to the uninitiated or novices in the group. This can be incredibly rewarding and fulfilling when done right.

-Trip Planner
This is an important one. There are many moving pieces when conducting an outing if a group is to be in the right place at the right time with the right gear to have a successful trip. This doesn't mean that the leader has to do everything, but they are responsible to see to it that everything gets done.

-Subject Expert
Training, experience, and good judgement are required for this type of leader. You don't have to be the best, but you may have the best "sense" about you. You might just be the go to person for advice on gear, navigation, first aid, weather, and much more.

-Coach/Mentor
Who here hasn't been helped through a difficult spot or encouraged when facing a difficult task. This type of leader helps folks overcome difficulties by adding encouragement and support to their base of knowledge. So many times the difficulty is in a lack of confidence. Having the ability to

positively assist someone to do their best and to see them succeed is one of the benefits of leadership.

-Guardian Angel
This is a difficult type of leadership. On a climbing team you are responsible for everyone's safety and well-being. You are the one making sure everyone has the right gear, right training, proper experience, fitness, and that the goal is within the grasp of the group. You learn to read the other members of your group and look for warning signs of possible risks to you and your group. I've had to be this guy and it isn't easy turning everyone around due to weather or route conditions.

-Arbitrator
In many group outings, there will come a time when there will be differences of opinion on many different matters. It is good to have group discussions on different aspects of an outing but when you have differences or tempers begin to flare this leader needs to be able to put their foot down, make a decision, and get the group moving in the right direction again.

Every leader will develop a different style and may be one or more of the different examples given above. You tend to develop your style through experience, through failure, through learning your craft and being effective in the backcountry. There is no exact way of doing this, but there are a few guidelines to help you on the way:

-A good leader cannot be self-centered; decisions are made for the benefit of the group.
-Having a genuine interest in every member of your group will foster a true caring of each other and will make you and your group stronger.
-Never pretend and do not show off. Be honest about your own abilities. If you don't know, say so and let the group help figure it out.
-Don't be afraid to laugh at yourself.

Safety:

Personal and group safety is paramount in the field. Every accident is different, but the contributing factors are very often the same. Inadequate experience and skill often lead to errors in judgement, poor tactics, and improper use of gear. This isn't saying that experienced folks don't have accidents, but they are able to mitigate the occurrence of them and lessen the severity of them. The Professional Citizen must place a high priority on individual and group safety and instill safe practice into the way they operate.

Setting an anchor for rappelling.

-Self Reliance
Most of us would recognize the responsibility for safety to rest with the individual first and that they understand the importance of being able to rely on oneself and avoid dependency on resources other than their own skills, knowledge, and equipment that they can carry with themselves into the field.

Reliance on things such as cell phones and GPS's, while nice and can speed up and simplify certain things, can often be an attempt to substitute for lack of preparation and self-sufficiency. I know there is a healthy debate on this, there has been for decades now, but it doesn't change the fact that you still need to have solid base skills first before supplementing with tech.

-Field Incidents
We try to avoid accidents; good safety practices mean studying and avoiding the broader scope of possible accidents. This would include such events as being lost, unplanned overnighters, negligent discharges, falls, frost bite, hyperthermia, animal bites, and many more.

Accidents don't just happen. Most reporting of accidents in the backcountry get attributed to human error. Even the most experienced amongst us have had accidents, and it is a fact that reminds us of how fragile life is.

-Hazards
Hazards fall into two categories, the most recognized are "objective hazards" which include physical dangers such as weather, darkness, high altitude, cliffs, glaciers/snow fields, rivers, and extremes of temperature.

Then you have "subjective hazards" which are the dangers created by people themselves that might be physically or mentally unprepared for the activities in front of them. Many times it is due to being out of shape, over confident in their abilities, lack of proper training, poor judgement, subpar gear, poor judgement, or even fear.

This book is filled with useful information to help you minimize the subjective hazards. Instead of inexperience and ignorance, you will be able to bring knowledge and skill in tackling the objective hazards. You will know the right kinds of gear to bring along with you, the right way to layer, how to procure water and even prioritizing needs in the field.

Today we have a special hazard as more and more citizens seek different avenues of preparedness. Folks are inundated with videos online, magazine stories, and even news stories and there is a movement for citizens to better prepare themselves. This causes an overabundance of folks heading to the local gun range for the first time, the nearby forests for their first campout, and people banding together. With this comes the need for better avenues of training and information sharing to help mitigate the subjective hazards, which in turn, helps to ease the objective hazards. The Professional Citizen Project was born out of this need.

-Risk
We, as individuals, need to manage risks. This isn't an adventure story with Arnold Schwarzenegger as the hero surviving 200 against 1 odds with nothing but a big knife and bulging muscles. We need to think our way through problems, all the while analyzing and evaluating, making decisions based on sound reasoning and leadership principles, all the while constantly looking for dangers and thinking ahead on what needs to be done to keep everyone safe while still pushing for a successful outing.

-Risk Assessment
I talk about risk assessment a lot when putting together trips and exercises. There is actually a formula to help in deciding on a plan of action. The formula looks like this:

Risk = Severity x Probability x Time

So, look at this as there being a more severe risk when there is an increase in any of the three variables listed above. The likely severity of an accident, the probability an accident could occur, and the length of time exposed to that risk.

Let's consider this, you're planning a 48-hour field exercise where you are patrolling a lengthy property line and setting up a temporary field camp. The weather forecast is for a 70% chance of rain with nighttime temperatures dropping into the 40's. Half your group is new to conducting foot patrols. You're thinking of sending the patrol out with bare bones sustainment kit. What is the risk assessment?

Let's work out the risk formula:
*Severity: A cold, wet night out in a bivouac could have serious consequences with minimal kit for staying warm and dry, especially for the new guys in the group. So, severity is high for hypothermia.
*Probability: It is unlikely you would fall to the cold and wetness through past experiences but the new guys in the group have never spent a night out under such conditions. Their inexperience increases the likelihood of succumbing to the conditions. This increases the probability of hyperthermia for someone in the group.
*Time: You will be out in the field for 48 hours so your exposure to the elements is high. Working in the field for the new guys will be tiresome. Time of exposure to the elements is high.

So, you can see, according to the formula, there is no simple answer but, it does point to a pretty obvious solution.

*Probable Solution: Have the guys take extra warm and wet weather gear and be prepared to have a hot camp.

By taking the time to look at the risks associated with any outing you can be better prepared and raise your chances of success.

Overall, the objectives for our risk assessments are to keep our perceived risks in line with the real risks, which helps to establish the amount of acceptable risk we are willing to take to achieve a goal or objective. This can be the culmination of all other safety measures including our gear, training, and experience along with the possible hazards regarding weather and terrain. When you understand the way these all work together, you will then gain a better understanding of acceptable risk and how much is right for you and your group. But don't let your judgement be clouded by your own overconfidence or wishful thinking, this can cause a dangerous level of risk that you may not really be ready to undertake.

Climbers on big mountains encounter this dangerous level of risk too often. They let their perceived risks get out of balance with the real risks and then accept that risk in a way that they normally wouldn't accept. They can be tempted by glory, by being the first, or from the standpoint of feeling that nothing bad happened the last time that level of risk was taken.

The moral of the story here with risk is to be able to pause and take inventory of the situation, of the people it involves, and to work out how best to achieve the desired result with the lowest chance of failure. I have developed a strong sense of a gut instinct that has served me well in the mountains. But a lot of that is due to experience. I've been lost, I've been cold, I've been late, and I have failed on far more peaks than I can list here. But it has helped develop a much stronger understanding of risk assessment and how to use it to breed success in the field.

First Aid:

Accidents can happen anywhere; they aren't just reserved for the backcountry traveler or soldier. However, The Professional Citizen must have the skills to manage the situation far from help. As we went over in the previous section, the more we can mitigate the risks in the field the more we can prevent accidents from happening, but they will still occur even in the best circumstances. Skillful prevention should be the goal for everyone in your group. This is much preferred to being an expert to treat injuries after they have happened.

However, every individual needs to be trained to a certain minimum in first aid. We can't just rely on the expert in the group only to have that person become the victim. This makes it so important that everyone be trained in first aid, not just the basics, but for use when more advanced care may be a day or more way in a remote setting. Please seek hands on professional training for first aid and CPR, *this section is not a substitute for training*, and strive to take refresher courses to stay on top of best practices.

-Outdoor Injuries

The outdoors presents a vast array of hazards, that are predictable in nature. Many of them are preventable, or if detected early, easily treatable.

Below are common outdoor-related injuries with a brief explanation of how they happen so that you know what to look for.

*Blisters: The dreaded blister, hated by outdoorsmen worldwide. These are probably the most common injuries for ruining many outdoor adventures. Blisters are often caused by new or poorly fitted boots. They result from your skin rubbing against your socks and the inner lining of your footwear. If your boots are too loose, too tight, or when your socks get wrinkled, or you get a small piece of debris inside your boot it can create a point of friction. Moisture also adds to this problem since it softens your skin promoting blisters to form.

The best prevention is having properly fitted and well broken in boots. The heal (and Achillies area) and toe areas are most prone to blisters so extra care is needed to ensure a proper fit. Keep your feet dry, wear socks made of wool or synthetic blends (no cotton), and, if prone to blisters, pad those spots with moleskin or tape to prevent the conditions that cause blisters to form.

*Dehydration: This is also a common ailment in the field. You lose water through breathing, sweating, urination, and diarrhea. A lot of times you don't even realize you are dehydrated until it is too late. We all lose water at different rates and we all have differences in our water intake needs.

Hydration is too important to be overlooked. Being well hydrated can reduce risks for heat and cold related issues and can have an overall impact on your physical performance in the field.

One aspect of good hydration practices is to be well hydrated before you even start off. Back in my mountaineering days we would begin hydrating anywhere from 3 days to upwards of a week before a big trip. Then when on the trip we would make sure to slowly drink throughout each day to maintain a certain level of hydration to stave off environmental effects. You do not want to wait until you are thirsty to drink, but rather drink before you reach that point, even if it's a cup or two every half hour to hour. If you don't feel the need to urinate or your urine is darker than normal, then you are not drinking enough. My son's Scout Troop had a cheer concerning drinking water:

"What do we want? Clear pee! How do we get it? Drink water!"

So, as you can see, it is really as simple as making sure you are drinking enough fluids to stave off dehydration and all the ailments that it can lead too.

*Heat Cramps: One of dehydration's biggest causes is heat cramps. During sustained physical exertion the body can cramp up as it detects an imbalance in hydration and electrolytes. Usually rest, muscle massage, and stretching along with hydration and taking in electrolytes can easily treat this.

*Heat Exhaustion: This is the milder of the two heat related illnesses (Heat Stroke being the worst). Heat exhaustion can happen when your body builds up more heat that it can lose, usually associated with extreme physical exertion or long exposure to a hot climate. As your body heats up, it makes an effort to reduce your body temperature by dilating your blood vessels that blood circulation to the brain and other organs begins to be reduced which, in turn, creates unhealthy blood levels. It feels similar to that moment when you feel lightheaded and faint. Symptoms can include feeling cold or clammy skin, faintness, weakness, nausea, and rapid heartbeat. Basic treatment can be rest, elevated feet, seek shade, and intake of fluids and electrolytes (see the trend of how bad dehydration is here)...

*Heat Stroke: Compared to heat exhaustion, this is now an emergency. Your body has gotten so hot at this point that you are in danger. Symptoms include confusion or unresponsiveness, rapid heartbeat, weakness, headache, and red-hot skin.

The mental state of the victim is the most visible symptom of heat stroke. Immediate treatment is needed, regardless of how uncooperative they may be. It is paramount that you start cooling the person. Seek shade, cool off the head and body with water or snow, and fanning. Constant monitoring is needed throughout the process, even once you get their temperature down it can easily go right back up with no notice as the body tries to recool itself. If the victim is able to, introduce cold liquids. Usually, a person suffering from heat stroke will have the trip over for them. In many cases, this means an end to the outing for al participants as you need to get this person to more advanced care.

*Sunburn: The effects of the sun are a year-round concern for us. Sunburn can be serious, but it is also easily preventable. Even on cloudy days UV radiation is not fully filtered out so prevention is still needed.

The best way to prevent sunburn is to cover exposed skin with clothing. We are lucky today that there are many options available to us for sun protective clothing. Add wide brimmed hats (think of the classic boonie hat) to cover your ears, face, and neck. Where skin remains exposed, then the application of a sunscreen product will help to alleviate the sun's effects on your skin.

*Hypothermia: Here we have the opposite of our heat related injuries. When you have lost more body heat than you are able to produce you are at risk for hypothermia. Just as with heat stroke, hypothermia is a medical emergency requiring immediate care and treatment.

Wet clothing and wind exposure lead to heat loss via evaporation, radiation, convection, and conduction. Dehydration (there it is again) can also be a contributing factor.

Most folks equate the onset of hypothermia to shivering caused by the body attempting to warm itself back up. When hypothermia gets worse, the shivering can actually stop. Common symptoms, in addition to shivering, include reduced fine motor skills, stumbling, slurred speech, decrease mental awareness, and even uncooperativeness. In severe cases the victim may even appear dead. This makes it essential to not give up on efforts to revive them and until they are warm when normal body functions begin to return.

Treatment options vary but the most common can be stopping the exposure to the cold and wind and remove any wet or cold clothing. Begin gentle rewarming by putting on dry and warm clothes, gentle massaging of extremities, and if able, introduce liquids for hydration. In extreme cases evacuation may be required.

Earlier we discussed risk and risk assessment, and in the example used we stated the patrol would see wet weather and temperatures in the 40's, this is a perfect example where deciding on preventative care to mitigate the chances of hypothermia are key to ensuring the group's overall safety and success.

Frostbite: The last example we will use is frostbite. This occurs when the actual blood vessels and surrounding tissue begin to freeze. Frost bite can cause permanent damage, and in many cases require amputation of the affected area. If caught early on, it can be reversed, but if left to get worse can be irreversible.

Treatment for frostbite begins the same as with hypothermia. Due to the nature of frostbite injuries, evacuation is usually best to get the victim to advanced care options.

But wait, there's more...

*Head, Back, and Neck Injuries: I mention this not only because of the commonality of such injuries in the field but also because it is something I have dealt with firsthand with my own injuries. I had an injury on a wilderness whitewater kayaking trip where the only way out was to continue down the river. It was a painful and agonizing experience that resulted in a slipped disk in my neck and a torn rotator cuff. I was not able to get to advanced care for 4 days and it made me hyper aware of the conditions and natural hazards that exist and just how careful you need to be, even when you do everything right, to navigate these potentially life-threatening injuries.

The head and spine are so delicate that even the smallest mistake in care may cause further injury. The biggest issue in care usually centers around whether or not you can safely move the victim. There are exceptions to this, but you will want to be sure to seek advanced training for conditions and treatment.

But lets end with common indicators of head injuries:

-Unconsciousness

-Slow pulse

-Fluctuating respirations

-Headache

-Drainage from ears, nose, or eyes

-Confused or disoriented

Time to Act
As an emergency unfolds, there are key things you must do in your response. We won't go super in depth, but the key actions are as follows.

1. Take Charge

2. Make sure it is safe to approach victim

3. Begin critical first aid

4. Treat for shock

5. Check for other injuries

6. Make a plan

7. Put the plan into action

First Aid Kits
It is everyone's responsibility to have a personal first aid kit. For The Professional Citizen we recommend 2 different kits. One is your blow out kit for plugging holes and stopping mass hemorrhaging and the other is a general "Boo-Boo kit" meant for general ailments and knicks and bruises.

First Aid kit examples

The best kits will be the ones you put together yourself according to the highest level of training you have received.

Your Blow Out kit should have a minimum of:
-Chest Seals
-Tourniquet
-Pressure Bandage
-Combat Gauze
-Rolled Gauze
-Nitrile Gloves

Your Boo-Boo kit should include, but not be limited to:
-Band Aids
-Moleskin
-Gauze Pads
-ACE Bandage
-SAM Splint
-Triple Antibiotic Ointment
-Hydrocortisone Cream
-Tweezers
-Ibuprofen, Antihistamine, Tums
-Tape
-Rolled Gauze
-Antiseptic wipes

Rescue:

We discussed pretty heavily the need for staying safe and healthy in the backcountry. But even the most prepared can experience catastrophic disaster and when this happens, we need to be prepared to evacuate to safety. Unfortunately, we can find ourselves in fairly remote locations where rescue may be hours or even days away. Most of us won't have QRF (Quick Reaction Force) or SRT (Search and Rescue Teams) to rely on when we need to make that call. So, this is something, as Mountaineers are proud of, The Professional Citizen must take a page out of their book and be proud of our own abilities to rescue our own.

Rescue skills are similar to first aid skills. They are learned skills that we hope to never have to use. But, when and if that time does come, you will be better prepared to deal with it. With this knowledge combined with experience you may be able to take a traumatic situation and turn it into a safe and successful experience for all involved.

So, let's focus on how we can use our skills, with our equipment, to mount such efforts and how we will need to be able to make such decisions.

What to do when an Accident Happens
This is probably the most stressful part of any rescue. The moment it happens you have to start deciding what to do. These initial decisions will ultimately decide the outcome.

My first recommendation is to be able to assess the situation. Make sure you and the rescue team are safe, take inventory of who is injured and to what extent the injuries are and begin treatment. If someone is lost, prepare to begin a search. Read through "Time to Act" in the previous section for a refresher on the steps to take.

Some groups may already have a designated first aid leader for directing medical efforts or an overall group leader in charge of the group. Many times, in emergencies, a natural leader steps forward to take charge, who knew you had it in you to lead!

When deciding the proper course of action consider these three things:

-The condition of the ill or injured, what treatment is needed, and is it safe to move them.

-The skill level of the others in the group. Does anyone else have rescue experience? Is the overall outdoors skill level high?

-What other hazards are present? How far is it to safety? What are the weather conditions like and how long till darkness sets in.

Rescue Rules
These are rules to abide by to help increase your chances of a successful rescue:

-First, consider the safety of the rest of your group. Their welfare comes before those needing rescue. It sucks to be in this situation but the last thing you want to do is make the situation worse by adding to the body count. No wis not the time for heroics.

-Second, do not rush things. We will all feel a certain spike in adrenaline but rushing into things leads to making bad decisions and mistakes soon follow. Remain calm, act with purpose, and be caring. The injured folks will notice this and help themselves remain calm.

-Third, Work within your scope of training and ability. Now is not the time to fake it.

Lost Person

This is a common rescue situation in the backcountry. Someone has either wandered off on their own or suffered a mishap where they became separated from the rest of the group.

Let's look at this from a case of a member of the group who just happened to get separated from everyone else and misses the rendezvous point. First, we should consider if they actually need help. If they are well equipped and trained, then maybe the best course of action is to simply wait until the next morning.

However, if dealing with bad weather, medical concerns, or dangerous terrain then you may want to mount a search. The most effective method would be to start at the last known spot where the missing person was seen. Try to imagine what errors they may have made at that point to get lost. There may be clues such as footprints, items left behind, or possibly trail markers they left to indicate the direction they headed.

If all else fails, and several hours of searching yield no luck, then it is probably time to look for outside help.

Evacuation

Sometimes we may have to deal with an injury such as a broken leg or busted knee that prevents the group members from walking out on their own. This is a very common situation in the backcountry. Someone slips on the trail and tweaks their knee rendering them unable to walk out on their own. If after a period of rest, they are still not able to continue then you need to put into action a plan for either assisting them out or to fully evacuate them. Be sure to work with the victim as to what they are comfortable with, monitor the injury, and always consider their mental well-being during the process.

Seeking Outside Help
There are times when the scope of the rescue is beyond your means. Outside help will need to be summoned for help. This isn't easy to do. Many times, help is a day or more away on foot. Options vary from sending a runner for help to calling in rescue via cellular, satellite, or radio communications. In good times all these methods are possibilities. Rescue beacons are a thing today and many Search and Rescue Teams get alerted to those beacons being activated. When all else fails then signaling will have to do. Whistles, fire, mirrors, Hi-Vis Panels, smoke, and more are methods for signaling for help.

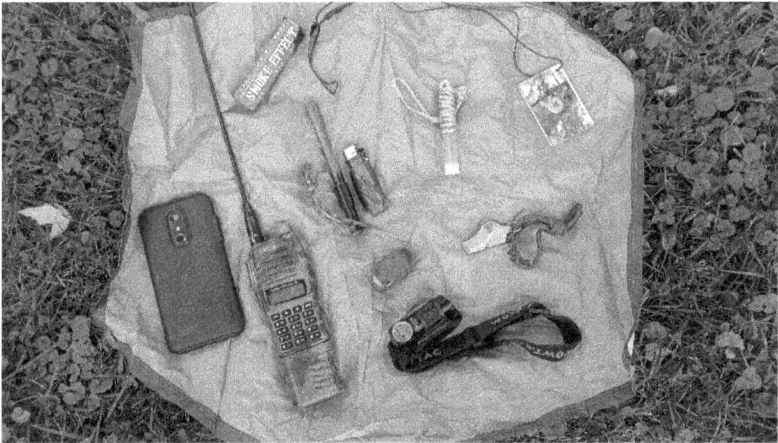

Example of my signaling kit.

I highly recommend the following for a signaling kit:

-Whistle
-Signaling Panel
-Colored Smoke
-Signal Mirror
-Buzzsaws (chemlight attached to a couple feet of cord)
-Lighter
-Radio with local emergency frequencies programmed
-Flares

Going for Help

If you need to send a runner for help, send two people together for safety reasons. Make sure they have a clear understanding of the situation and what help is needed. Your runners should be carrying the gear they need to get them safely to their destination, but not so much that they can't do it swiftly.

Their priority is getting to help safely and to make contact with the appropriate authorities to request help. They must be able to accurately report what is needed such as location, special circumstances, number of people needing help, and to what extent the injuries are.

Sample Emergency form

HIKING PLAN

Complete this form before departing on a hike and leave it with a reliable person who can be depended upon to notify authorities in case you do not return as scheduled. A word of caution: in case you are delayed and it is not an emergency, inform those with your hiking plan of your delay in order to avoid an unnecessary search!

1 Names of person filing this plan: _____
Telephone #: ( )

2 Name(s) of others on hike: Age: Address:
Telephone #: ( )
Telephone #: ( )
Telephone #: ( )
Telephone #: ( )
Telephone #: ( )
Telephone #: ( )

3 Radio ☐ Yes ☐ No   Type: _____   Call sign: _____   Frequencies: _____

4 Trip plans
Leaving from: _____   Going to: _____
Route details: _____

Departing on: ___/___/___   ___ ☐ am ☐ pm   Returning: ___/___/___   ___ ☐ am ☐ pm
         date   time   date   time
And, in no event, returning later than: ___/___/___   ___ ☐ am ☐ pm
                                      date   time

5 Alternate route if bad weather is encountered : _____

6 Description of automobile: _____
Make: _____   Color: _____   License #: _____   Where parked: _____

7 If not returned by: ___/___   ___ ☐ am ☐ pm
                      date   time
Call: _____
Local authority: _____   Telephone #: ( ) _____

89

Notes:

CHAPTER 3
Surviving Your Environment

Understanding your environment is crucial to working, and surviving, in the best and worst of times. While we can't layout everything for surviving in the mountains, jungles, deserts, and everywhere in between (there are plenty of books that go into great detail for those areas), we can introduce basic survival concepts that can be used just about anywhere to help increase your odds of survival and rescue.

S Size Up the Situation
(Surroundings, Physical Condition, Equipment)

U Use All Your Senses,
Undue Haste Makes Waste

R Remember Where You Are

V Vanquish Fear and Panic

I Improvise

V Value Living

A Act Like the Natives

L Live by Your Wits, *But for Now,* Learn Basic Skills

Acronyms are useful memory tools for remembering important information.

Psychology of Survival:

It takes much more than the knowledge and skill to build shelters, gather food, build fires, and move without the aid of standard navigational devices, to survive alone and possibly in detention or captivity. The key ingredient in any survival situation is the mental attitude of the person involved. Having survival skills is important, having the will to survive is essential. Without a desire to survive, acquired skills serve little purpose and valuable knowledge goes to waste.

During any survival situation, folks will experience numerous stressful moments that will shape them for the remainder of their lives. Stress is not a disease that you cure and eliminate. Instead, we all experience this condition. Stress is our reaction to pressure. It is the name given to the experience we have as we physically, mentally, emotionally, and spiritually respond to life's tensions. Who hasn't experienced stressful situations in their lives? How have you dealt with it in the past? You can apply those experiences to a survival situation.

Prepare Yourself Now
Your mission in a survival situation is to stay alive, period. The mixture of thoughts and emotions you will experience in a survival situation can work for you, or against you. Fear, anxiety, anger, frustration, guilt, depression, and loneliness are all possible reactions to the many emotions common in survival situations. These reactions, when managed in a healthy way, will help you to increase the likelihood of surviving. They prompt people to pay more attention in training, to fight back when scared, to take actions that ensure sustenance and security, to keep faith with their fellow team members, and to strive against large odds. When folks cannot control these reactions in a healthy way, they can bring themselves to a standstill. Instead of rallying their internal resources, they listen to their internal fears, which causes the isolated person to

experience psychological defeat long before they physically succumb. Remember, survival is natural to everyone; being unexpectedly thrust into the life-or-death struggle of survival is not.

Be Realistic
Folks should not be afraid to make an honest appraisal of their situation. They should see the circumstances as they are, not as they want them to be, while keeping hopes and expectations within the scope of the situation. When going into a survival setting with unrealistic expectations, one may be laying the groundwork for bitter disappointment. Follow the old adage, "Hope for the best, prepare for the worst." It is much easier to adjust to pleasant surprises about unexpected, good fortunes than when confronted with unexpected harsh realities.

PMA (Positive Mental Attitude)
Your attitude in a survival situation can ultimately decide the outcome. It affects everything you do, every decision you make, and can affect those around you.

Now is the Time to Train
There is no better time than now to begin your journey to prepare to handle the difficulties of survival. Folks capable of performing their skills in training today will have the confidence in their ability, when called upon, to perform them under distress. The overall goal of preparing now is to build confidence in your ability to function despite your situation and fears. Failure to prepare yourself physically, mentally, and emotionally to cope with survival leads to reactions such as depression, carelessness, inattention, loss of confidence, and poor decision-making. Remember that your life and the lives of others who depend on you are at stake.

Priorities:

In any survival situation, there are key priorities that need to be met to give yourself the best chance at survival and rescue. The basic four priorities of survival would be, in order:

-Shelter
-Water
-Fire
-Food

I would go as far as to add a fifth priority and that would be signaling for rescue. Surviving does you no good unless you can be rescued as well.

So, Let's explore each of these survival priorities in a bit more detail.

Shelter
Protection from the elements is crucial in the outdoors. Getting out of the wind, rain, snow, cold, sun, away from insects, and even enemy observation are all important to your long-term survival. It can give you security, a feeling of wellbeing, and even help you to maintain your will to survive.

In many cases your initial shelter from the elements will be your clothing, or uniform. If you go back and read through the earlier section on clothing and the layering system, you will find that your clothing choices can, and will, affect your chances of survival. Your clothing needs to be able to keep you warm, dry, cool, out of the sun and away from insects.

For any situation that requires shelter, the first thing to do is to assess your situation and the items you have at hand for making your shelter. Are you in a location safe from fire, flooding, and clear of anything that could collapse on

94

you. Your shelter doesn't have to be a castle, but it must be able to keep you comfortable. Tarp shelters, snow caves, debris huts, and natural formations are all great shelter ideas and require practice during training to get it right when they really count. Keep in mind that you will be limited by calories, time, and ability. You don't want to dedicate precious energy and time to a shelter build when you still have other survival priorities to consider.

-Tarp Shelters: Probably the easiest shelter to practice and become an expert with is a tarp shelter.

Basic plow point shelter utilizing a poncho.

Tarps can be many different sizes, from simple ponchos to heavy duty canvas tarps. There are a myriad of ways for setting them up from simple plow point shelters (as seen in the photo above, to your classic A-frame setups to even more elaborate tarp tents. Many times, weather will determine the best type of setup, keeping in mind the need to best protect you from the elements.

In my own kit my basic shelter setup consists of an old army poncho, some paracord, and a few tent stakes.

Through experience and training, I know that this basic kit enables me to be able to setup a shelter quickly, in a variety of ways, with little energy expelled so that I can move on to other needs.

TARP SHELTERS
WITH ONE TARP AND SOME PARACORD

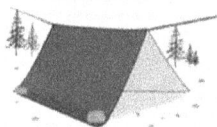

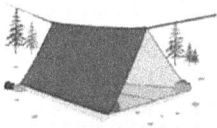
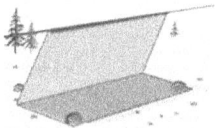

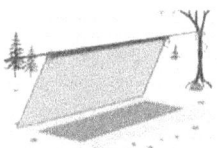
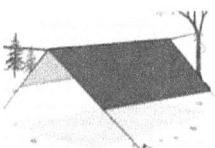

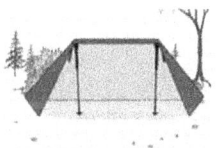
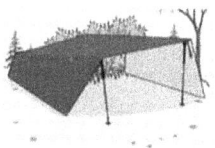
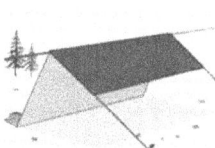

Examples of various tarp shelters.

-Snow Shelter: While tarp shelters are nice, they lack insulation qualities to them to provide you better cold weather protection. I live in an environment where winter is a real season and for a few months of the year survival in it can be a necessity. A basic shelter to practice making and using is called the Quinzee. In wet, snowy conditions, the Quinzee offers protection for either a single person or a small group.

Quinzee constructed for one person.

1. For building a Quinzee, place your gear in a pile and cover with a tarp (if you have one).

2. Begin piling snow on top of your gear and continually patting it down. Insert sticks (should be 12" to 18" long) around the mound to act as depth gauges and allow the snow to harden for about an hour.

3. Dig out a doorway, pull out your gear, and further excavate the inside until you hit the depth sticks you had inserted to prevent thin spots in the dome.

4. Add a door for further protection and you're ready to occupy it for the night.

Building a Quinzee

One
Pile snow into a large mound. Piling snow on back packs helps create the internal space that you need to make

Two
Push sticks 6 to 8 inch into quinzee. This measures the thickness of the walls

Three
Carve out quinzee from both ends

Four
Continue to carve out the inside, paying attention to the roof thickness.

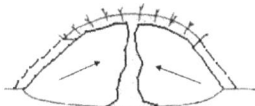

Example of building a Quinzee.

-Debris Hut Shelter: A third type of shelter to learn to create is one that uses the natural surroundings in the form of a debris shelter. Relatively quick to make using downed branches, sticks, logs, and leaves, these can be built in minimal time with little effort. Simply constructed using a single support beam as the main support with branches leaned against it on both sides creating a structure very similar to the plow point tarp shelter mentioned above.

Example of a debris shelter.

You can then cover it with leaves, pine boughs, a tarp, even snow to protect, and even insulate, you from the elements.

Water
We can only live about 3 days without water. Being able to source water on day one of survival is vital to your overall success. There is a reason water, and the means for procuring water, is amongst our ten essentials listed in Chapter 1. It is an absolute necessity for life and in the filed

we have to be able to put in the work for every sip we take. Do not take water for granted.

The first step in your water needs is to have a plan. Sounds simple, right, but you would be surprised how many folks take water for granted. Most of us grab our water bottle as we head out the door for the day, but just how far in advance are we planning? Let's go over some basic tips to help ease the burden of running out of water.

-Bring it with you: It is a good idea to bring plenty of water with you. If you are in a water rich environment (I live in an area with many creeks, rivers, ponds, and lakes) you can get away with less but in more arid climates you may need to plan on hauling a few days' worth of water with you. Will it be heavy, sure, but the effort is worth it to have readily available water.

-Treatment: Water filters, purifiers, iodine and chlorine tablets, boiling, and even UV treatments can all make questionable water safe to drink. Just as with other gear, you should be carrying multiple methods for water treatment.

-Always be Scanning: As you are out always keep your eyes open for water sources to draw from. Your map may show a water source, but it could be dried up. Be prepared for unexpected water source such as muddy holes, vines, depressions in rocks, and so many more. In a survival situation we can't be picky about where the water comes from, treat it and drink it.

Weather and environment can also determine your water source. Rainfall is easy, snow takes a bit of work to melt, and morning frost/dew is easily collected with a rag or shirt.

Plants are another great source for water. Trees such as Maples, Birch, Hickory, and Sycamores (one of many reasons tree and plant identification are important) all can be tapped for water. Simply cut a notch into the tree, insert a small stick into it, and set a container underneath to collect it. It isn't fast but it is a reliable method for water collection in the woods.

Safe drinking water will make or break you. I know we see it all the time on "survival" shows where the contestants become ill and have to "tap out" due to drinking unsafe water. Time and time again it is the water that does them in. Well, in a real survival situation, we can't just push a button when we make that mistake. We have to live with it, or not, depending on how you want to look at it. Here are the hidden dangers we are trying to protect you from.

*Viruses: These are the smallest of all the water borne pathogens that can get us. They are small enough to even infect bacteria present in the water. These waterborne viruses include hepatitis A, rotavirus, norovirus, and enterovirus.
*Bacteria: Pathogenic bacteria often enter our water ways through dead wildlife and feces in the water. Plenty occur naturally as well in our surface water. Disease causing bacteria that can be transmitted include cholera, Salmonella, and Staphylococcus.
*Protozoa: This would include Giardia and Cryptosporidium as the common forms of pathogens present in untreated water. Giardia is the nasty one that sneaks up on you that'll give you diarrhea and requires about $300 worth of medication to get rid of. You do not want to get Giardia. It sucks, believe me.

Fire
The means and skill to make fire is essential for boiling water, cooking, heat, light, signaling, and safety. We discussed fire in Chapter 1 but let's explore a couple of fire-starting methods and uses when in a survival situation.

When I first started getting serious in the outdoors the rule of thumb was that you kept a lighter wrapped in duct tape in every layer. Pants pocket, jacket pocket, pack pocket, fire kit, stove kit, and survival kit. Basically, the idea was that no matter what you had on you would have a means for starting fire. Well not much has changed with this methodology except maybe the ignition sources we use. Plenty of folks still use lighters, I know that I still do, but I have also broken my fair share of them without realizing it. So now it is becoming common for folks to keep ferro rods in their different pieces of kit. Someone well practiced with a ferro rod can start a fire just about anywhere. Just don't lose it like that one guy on Alone. Remember, no tapping out in a real survival situation. I'm a fan of the rule of three's when it comes to fire so a lighter, waterproof matches, and a ferro rod round out my fire making ignition methods. So, the main uses for fire in a survival situation would be for warmth, signaling, safety, boiling water, cooking, and light. Fire is truly life in the backcountry.

-Fire uses besides warmth:

Signaling is important in a survival situation, you can do everything else but if you're not prepared to signal for help, then your survival situation might as well be permanent. The most common signaling fire would be a smoke generator.

Example of a smoke generating signal fire.

Key factors in the effectiveness of signaling fires is to have them ready to light at moment's notice and for them to be in the open to be easily seen. Beaches, pastures, clear cuts, and rock outcroppings are great locations for these. Another consideration is their size, the bigger the better and the more green materials used on top then the thicker the smoke plume will be. These can be seen from far off distances and once lit can be fed for a long time.

Boiling of Water means having safe drinking water. A metal container makes the boiling of your water easy and is reusable. The general rule of thumb is to boil your water for three minutes. Now that is three minutes of a rolling boil to be safe. Now this same fire can be used to melt snow for drinking water in winter conditions. It is time consuming but a necessity when faced with a survival situation.

Cooking can also be a necessity when processing wild game in the field. Maybe you are using snares and dead falls to try to catch game. You might even have a .22 kit gun at your disposal for dispatching small game. Regardless you will need to cook it in the field and the ability to make fire allows you the ability to eat. Every calorie counts in the field so being able to replace those lost calories is key to any survival situation.

Finally, a bonus tip for fire making...
-If you are a hunter or a true Professional Citizen, chances are you have a firearm on you which means you have ammunition as well. Great because the powder from your ammo can be used in your fire making strategy. Whether you're using a ferro rod, lighter, or a friction fire method, powder will catch a spark and burn hot and fast.

Food

In a multiday survival situation, you will probably spend a great deal of time looking for food. Calories are king here and every little bit matters. Plants may play a big role in your overall diet, but they will not be able to provide the bulk of your caloric needs. That's why it is so important to pack rations appropriately and to learn how to trap, fish, and hunt.

Unless you're a pro at identifying wild edibles, I would stay away from them and focus on small game and fishing. But lets start with the food you bring with you.

We discussed food in chapter 1 but lets briefly revisit that. As a rule of thumb, pack enough food for an extra day in the field. Even if you are only planning to be gone a few hours, throw a days' worth of rations in your pack just in case. It is better to have it and not need it than to need it and not have it, right. The prepackaged ration packs I highlighted are easy to put together and store. There is no excuse to not have enough food on you when on a trip. However, accidents happen and we might have to survive beyond what the supplies we brought with us.

So, lets look at some common tools you can use to actively hunt food while you can tend to other survival matters.

-Snares are an effective way to catch small game. They are easy to set, and you can set dozens of them to check on daily.

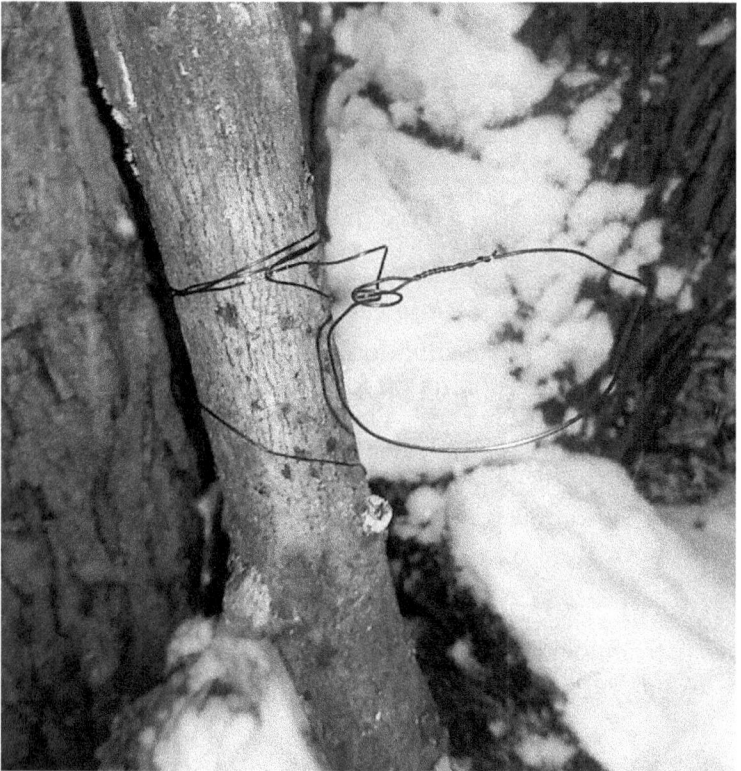

Wire snare set on a squirrel pole.

In addition to wire snares, you can fashion trigger snares from paracord with a trigger device. These take a bit of practice but when done right, they are effective along small game trails. Again, set several of these to actively work for you while you are tending to other survival needs.

Example of a spring snare

The idea here with traps is to start establishing a sort of trap line that you can work on every day. Setting these lines is a common practice for folks living in the backcountry that make, not only their livelihood from trapping, but also do so you stock their freezers with meat for the year. Setting and checking these lines becomes a daily occurrence but creates a system of actively working for you day in and day out as you see to your other chores. The more traps you can set early on the greater your chances are for catching something.

Variety in traps is good too. Different methods work in different areas. You do not want to put all your eggs in one basket, so to speak. So, let's look at one more trap before moving on.

-Deadfalls are another example of a small game trap that, once mastered, can be a very effective tool in catching small game.

Example of a Figure 4 Deadfall trap

The simplicity of the Figure 4 Deadfall makes it a favorite of mine. It can be built quickly with a few sticks and a sharp knife and can be reset time and time again.

Fishing is another relatively easy means of food procurement. The hardest part will be bait choices and patience. A small fishing kit consisting of monofilament line, small weights, and an assortment of hooks. With this simple kit you can catch plenty of fish.

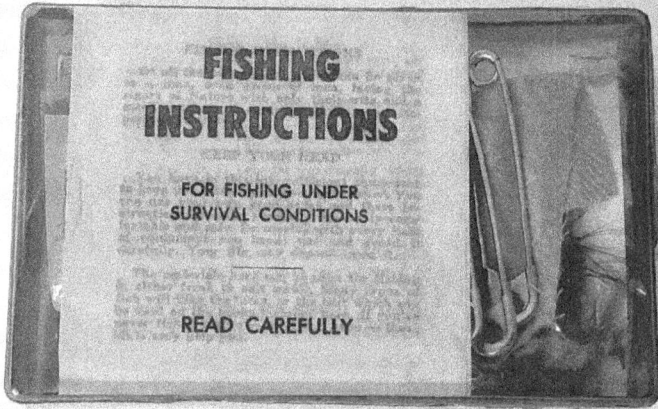

Example of a military fishing kit

As long as you have an assortment of hooks, line, and weights, then adding a fishing pole to your setup is all that's left. Here are 3 easy steps for doing so:

1. Find a flexible sapling about 1" thickness at the base and anywhere from 6' to 9' long. These skinny trees are strong and flexible.
2. Cut the tree with whatever tool you have.
3. Trim off al the branches and tie your line to the end and you are ready to start fishing.

Signaling
As we discussed earlier, signaling is key to being rescued. While it may not be as much fun to practice as, let's say, fire making or trap building, it is the way you are getting home. Whistles, signaling mirrors, and a smokey fire are the 3 most common ways for signaling for help. But let's look at a couple of other ways to help you in being found.

-Ground-to-air signal construction is a great way to get a large, visible signal that once constructed, can be left while you attend to other needs. You want to be sure that it is in an open area and large enough to be spotted by a passing aircraft. Here are some common ground-to-air signals and their meanings:

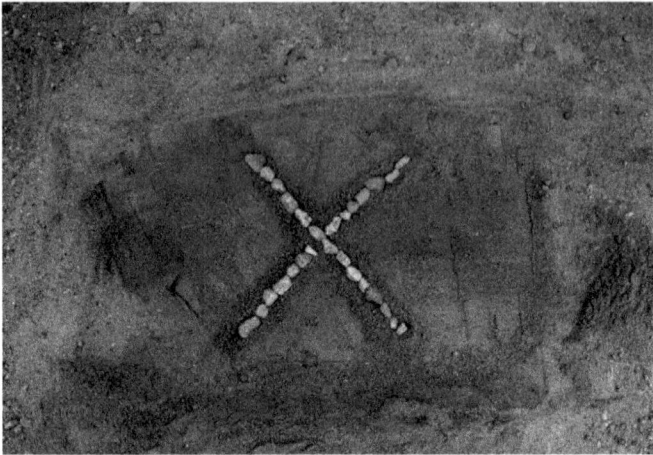

An "X" made from rocks

*A giant "V" shape is the international signal for "we require assistance". But it could be mistaken for a naturally occurring shape. But it can be easily built with logs or other long materials.
*A giant "X" means we require medical assistance or are unable to continue. It isn't as frequent of a natural shape in the backcountry. It is easily constructed with logs, rocks, or dug into soft ground.
*You can also spell out "SOS' or "HELP" using the same methods as above.

SOS stomped out in the snow

Ground-to-air signals actually have an effective track record of saving lives. But to give yourself the best chance of them working, here are some basic rules to follow.

*Get to open ground. A signal made under a canopy of trees is not going to help you. Get out in the open or find a clearing to use, if you are near water find a beach. Give yourself the best chance possible.
*Make it BIG. Think crop circles big. You want your signal to not only catch the attention of folks looking for you, but also those not looking for you.
*Make it contrast with the background. If on snow or light-colored sand, use darker colors. Pine boughs, burnt logs, charcoal, palm leaves, etc. If in a dark area use lighter colored items. Rocks, sun bleached logs, really anything that will stand out to passersby.
*Make sure it doesn't appear like a natural occurrence. It needs to look man-made and out of place to be noticed.

-Chemlight Buzzsaws are an effective means of signaling at night. Chemlights are simple, lightweight, and cheap to buy and take up little space in your kit. Not only can they be used as an emergency light source, but they also make for a great nighttime survival distress signal.

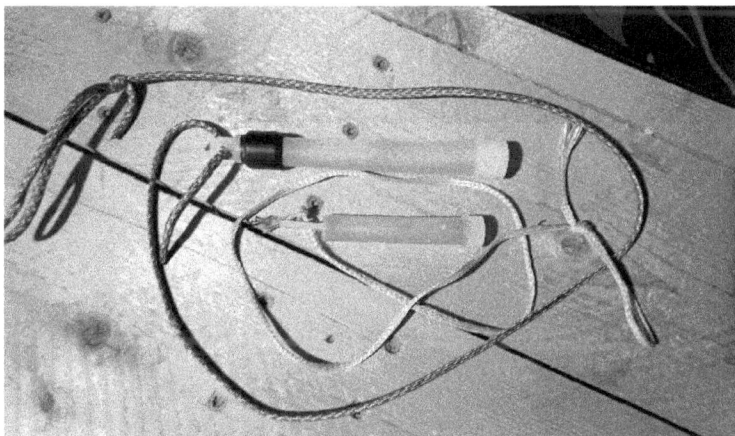

Examples of chemlights setup for buzzsaws

110

A chemlight on its own will not be enough though. It needs to be used in a fashion that attracts attention to it from a distance. To make the Chemlight Buzzsaw effective, swing your glow stick in circles as fast as you can with the circles facing the direction of where you want to signal. This creates a very unnatural full circle of light. It is unmistakable in a nighttime setting. It can be confused for nothing in nature and very little even in urban environments. It is an excellent signaling method for self-marking by casualties or ground to air signalling.

Buzzsaw being used to signal a helicoptor

Movement & Navigation:

When you are faced with a survival situation, the ability to move effectively is important to your rescue as well as to your safety and the conserving of energy and resources. Equally important is the ability to perform land navigation. The ability to determine location and navigate cross-country significantly increases your chances for rescue. Fundamental to land navigation is the acquisition of detailed knowledge about the environment you are in including the climate, terrain, proximately to civilization, and more. This section will provide an overview of movement and navigation considerations.

Do You Stay or Go
The decision to stay in place or to move while in a survival situation requires careful consideration. The reasons to stay in place or move are typically based upon the following:

*The current location does not provide adequate food, water, shelter, or the ability to signal for search and rescue efforts.
*Injuries or circumstances (such as physical or mental considerations) that would prevent you from moving your location.
*The current location does not provide multiple escape routes (SERE) and/or contains numerous danger areas that make staying put dangerous to you and your group.
*Your plans require you to meet up at a certain rally point for extraction or resupply.
*Enemy situation. Does the enemy composition, disposition and activity threaten the isolated person's ability to evade or survive?
*You are in an environment where military and/or law enforcement agencies have control as well as the intent and capability to assist recovery operations may favor staying near the survival site (ex. aircraft crash) initially.
Movement is considered only when certain that water,

shelter, food, and help can be reached, or after having waited several days, they are convinced that. recovery is not coming, and they are equipped to travel.

Other factors that should be considered prior to deciding to stay put or move include the following:

*Do you or any group members have a head injury or condition that affects their ability to make a clear decision? Avoid making any decision immediately after a survival situation presents itself. If possible, wait a period to allow for recovery from the mental (if not the physical) shock resulting from the event. When shock has subsided, evaluate the situation, analyze the factors involved, and make valid decisions. Better to be safe than sorry.
*Are the necessary tools, gear, and materials available to support movement? Movement can be risky unless the necessities of survival are available. To leave a safe shelter to travel in adverse weather conditions is foolish unless in an escape or evasion situation.

Once the decision is made to move, a couple considerations apply regardless of the situation:

*Continue to assess your physical abilities (rest, hydration level, caloric intake and output, injury prevention, pace and durability, mental and emotional state). Stop, think, and act before a problem arises.
*In a group situation, the acting leader must assume leadership and direct the efforts of the team during movement and rescue.

MOVEMENT CONSIDERATIONS
Proper movement maximizes speed and mitigates energy output along a route. The best posture when moving should balance the person's body weight directly over their feet. Dependent upon the requirement to move stealthily— the soles of the feet should be flat on the ground. Step over obstacles and not directly on them as they may pose a

serious tripping hazard. When ascending a terrain feature, lock your knees with each step to assist the joints in carrying their weight. Traverse or use a zigzag motion moving up steep slopes to conserve energy and help maintain balance.

When descending a terrain feature, keep your back straight and keep your knees bent so they do not overextend the knees joint. The traverse is also used to descend steep terrain. After determining the step is stable, transfer weight from the lower foot to the upper foot and repeat the process. Additional considerations include:

*Using game trails when they follow a projected course only in addition to an easier route of travel and the chance of securing game or locating water.
*Surveying the surrounding countryside, and plan movement only after carefully surveying the terrain.
*Studying your back trail carefully. Know your route backward equally as well as your route forward.
*Continually assessing the climate and geography as you move. Asking yourself questions such as:
    -Is the weather changing?
    -Is the route leading you into an area that does not support food, water, or shelter?
    -Does the movement require a change in survival techniques?
*Making camp early so that you have plenty of time to build a shelter.
*Using the buddy system to watch for heat and cold injury.
*Do not placing your hands or feet anywhere without first looking to see what is there.
*Grazing type animals (including deer) are better indicators of water than predators/carnivores.
*Bees seldom range more than four miles from their nests or hives. They will usually have a water source in this range. Ants need water. A column of ants marching up a tree is going to a small reservoir of trapped water.

Mountainous and Cold Movement

When conducting movement in a mountainous or cold environment, The Professional Citizen must consider the following:

Alpine conditions on a winter trip

*Avoid possible avalanches of earth, rock, and snow, as well as deep crevices in ice fields.

*Movement on the wind packed side of a ridge is typically more advantageous because the snow surface is typically firmer and there is a better view of the route from above.

*A loose snow layer underneath is more hazardous than a compacted one as the upper layer of snow
will slide more easily with no rough texture to restrain it.

*Leeward slopes collect snow that has been blown from the windward sides, forming slabs or sluffs, depending upon the temperature and moisture.

*Use a pole to probe ice and snow conditions during movement

*Avoid crossing glacial areas during the day.

*Avoid traveling during a blizzard.

*Take care when crossing thin ice. Distribute your weight by lying flat and crawling.

*Cross-streams where the water level is lowest.
*Wind-chill is a factor in all activities. Personal movement generates wind-chill that can lead to increased cold weather injury.
*Have enough clothing to protect from the cold; know how to maximize its warmth. For example, always keep your head covered. An isolated person will lose 40 to 45 percent of their body heat from an unprotected head and even more from the unprotected neck, wrist, and ankles.
*The brain is very susceptible to cold and can stand the least amount of cooling. Because there is much blood circulation in the head, most of which is on the surface, you can lose heat quickly if you do not cover your head.

Mountain Terrain Levels
Mountain outings and exercises are generally carried out at three different terrain levels. (Levels I, II, and III with more detail.)

Level I terrain is located at the bottom of valleys and along the main lines of communications. At this level, mounted groups can operate, but maneuvering space is often restricted. Dismounted and mounted groups are normally combined, since vital logistic lines usually follow the valley highways, roads, and trails. Most, but not all, of the civilian population is found at this level.

Level II terrain lies between valleys and shoulders of mountains. Generally, narrow roads and trails, which serve as secondary lines of travel, cross this ridge system. Ground mobility is difficult. Additionally, since dismounted groups can easily influence operations at level I from level II, they often expend great effort on these ridges. Similarly, enemy positions at the next level can threaten work on these ridges. The enemy can often find sanctuary at this level in the form of bunkers and caves. Ahem, Afghanistan anyone...

Level III includes the dominant terrain of summit regions. Although summit regions may contain relatively gentle terrain, mobility in level III is usually the most difficult to achieve and maintain. Level III terrain can provide opportunities for well-trained groups and individuals to attack the enemy from the flanks and rear. At this terrain level, acclimatized groups with proper skills and equipment can infiltrate to attack lines of communication, logistics bases, air defense sites, and command and control facilities.

Examples of the 3 levels of mountain terrain

EFFECTS ON OPERATIONS
Mountain terrain and weather affect nearly every aspect of mountain operations. The physical characteristics of mountains:

*Affect mobility and lengthen movement times.
*Physically and mentally taxing on folks
*Affect the operation and accuracy of some weapons.
*Challenge sustainment operations.
*Create hazards and risks.
*Complicated medical evacuation and casualty evacuation
*Interfere with line-of-sight radio communications.
*Challenging command and control.

In preparation for an outing/exercise, leaders should consider the following list of specialized activities, procedures, and techniques that may be required for successful trips into mountain terrain:

*Basic climbing techniques.
*Mountain navigation.
*Mountain stream crossing.
*Mountain terrain route selection.
*Off-road and steep-terrain driving.
*Use of ropes
*Procedures to avoid landslides and avalanches.
*Use of animal transport logistical items.
*Walking and movement techniques for steep and rough terrain.
*Cold weather movement (snowshoe, ski, sled operations).
*Mountain survival techniques.
*Hazardous cross country night movement.
*Advanced first aid.
*Personal hygiene and field sanitation.
*Small group standard operating procedure (SOP) and immediate action drills.
*Rough terrain/steep earth

Different mountain chains will have different types of climates. Some chains are in dry desert regions with temperatures ranging from extreme heat in the summer to extreme cold in the winter. In tropical regions, small to medium mountains are covered in lush jungles with deep ravines that flood during the rainy season. Temperatures in these areas typically remain warm and humid all year. Many of the mountains in Central America and many mountainous regions in Africa and South America that are located close to the equator have these characteristics. Conversely, high mountains in temperate climates have sparse vegetation at elevations above 3,505 meters (11,500 feet) and temperatures drop below freezing in winter. Some mountainous regions have a variety of environments, such as in Afghanistan where units have encountered several different mountainous environments within the same area of operations.

Mountains may rise abruptly from the plains to form a giant barrier or ascend gradually as a series of parallel ridges extending unbroken for great distances. Mountains may have isolated peaks, rounded crests, eroded ridges, and high plains and be cut by valleys, gorges, and deep ravines. High rocky crags with glaciated peaks and year-round snow cover exist in mountain ranges at most latitudes along the western portion of the Americas and in Asia. Regardless of their appearance, rugged terrain is common among all types of mountains.

Mountain slopes generally vary between 15 and 45 degrees. Cliffs and other rocky precipices may be near vertical or even overhanging. Aside from obvious rock formations and other local vegetation characteristics, actual slope surfaces are relatively firm earth or grass. Grassy slopes may include grassy clumps known as tussocks; short alpine grasses; or tundra, which is more common at higher elevations and latitudes. Many slopes will be scattered with rocky debris deposited from the higher peaks and ridges. Extensive rock or boulder fields are known as talus. Slopes covered with smaller rocks, usually fist-sized or smaller, are called scree fields. Slopes covered in talus are often an easy ascent route. On the other hand, climbing a scree slope is difficult because the small rocks tend to loosen easily and give way.

In winter and at higher elevations throughout the year, snow covers slopes, creating an environment with its own distinct effects. Some snow conditions aid travel by covering rough terrain with a consistent surface. Deep snow, however, impedes movement and requires folks to be well trained in using snowshoes, skis, and over-the-snow vehicles. Steep, snow-covered terrain presents the risk of snow avalanches as well. Snow can pose a serious threat to The Professional Citizen not properly trained and equipped for movement under such conditions.

COLD WEATHER CHARACTERISTICS
Cold weather conditions can be classified using categories. The temperature categories are:
*Wet cold, +39° F to +20° F (4° C to -7° C).
*Dry cold, +19° F to -4° F (-7° C to -20° C).
*Intense cold, -5° F to -2° F (-20° C to -32° C).
*Extreme cold, -25° F to -40° F (-32° C to -40° C).
*Hazardous cold, -40° F (-40° C) and below.

-Wet cold conditions occur when wet snow and rain often accompany wet cold conditions. This type of environment is more dangerous to folks and equipment than the colder, dry cold environments because the ground becomes slushy and muddy and clothing and equipment becomes perpetually wet and damp. Because water conducts heat 25 times faster than air, core body temperatures drop if folks are wet and the wind is blowing. The Professional Citizen can become a casualty due to weather if not properly equipped, trained, and led. Wet cold environments combined with wind is dangerous because of the wind's effect on the body's perceived temperature. Wet cold leads to hypothermia, frost bite, and trench foot. Wet cold conditions are not only found in mountain environments but in many other environments during seasonal transition periods. Under wet cold conditions, the ground alternates between freezing and thawing because the temperatures fluctuate above and below the freezing point. This makes planning problematic. For example, areas that are trafficable when frozen could become severely restricted if the ground thaws.

-Dry cold conditions are easier to live in than wet cold conditions. Like in wet cold conditions, proper equipment, training, and leadership are critical to successful operations. Wind chill is a complicated factor in this type of cold. The dry cold environment is the easiest of the four cold weather categories to survive in because of low humidity and the ground remaining frozen. As a result, people and equipment are not subject to the effects of the thawing and freezing cycle, and precipitation is generally in the form of dry snow.

- Intense cold can affect the mind as much as the body. Simple tasks take longer and require more effort than in warmer temperatures and the quality of work degrades as attention-to-detail diminishes. Clothing becomes bulkier to compensate for the cold, so folks lose dexterity.

- Extreme cold increases the challenge of survival becomes paramount. During extreme cold conditions, it is easy for folks to prioritize physical comfort above all else.
People will withdraw into themselves and adopt a cocoon-like existence. Leaders must expect and plan for weapons, vehicles, and munitions failures in this environment. As in other categories, leadership, training, and specialized equipment is critical to the ability to operate successfully.

- In hazardous cold conditions, leaders and planners assume greater risk if they engage in operations when the temperature falls below -40° F (-40° C). Groups must be extensively trained before undertaking any field exercise or outing in these temperature extremes.

Desert Movement
When conducting movement in a desert environment, evaders must consider the following:

American Southwest arid desert conditions

*Avoid salt marshes. Water in these areas is typically undrinkable without significant purification effort. The area is highly corrosive to skin, equipment and clothing.
*Expect a large thermal shift between day and night. The drop in temperature at night occurs rapidly and will chill a person who lacks appropriate clothing.
*Protect radios and batteries from direct sunlight while moving in the desert environment
*Rest during the day, work during the cool evenings and nights.
*Hide or seek shelter in dry washes (wadis) with thicker growths of vegetation and cover from oblique observation.
*Use the shadows cast from brush, rocks, or outcroppings. The temperature in shaded areas will be cooler than the air temperature in exposed areas.
*Use the 1:3 rule when judging distance in the desert. What appears to be 1 kilometer away is really 3 kilometers away.
*Expect major sand and/or dust storms at least once a week. To avoid becoming lost, do not move during these storms.
*Mirage makes it difficult to identify targets, estimate range, and see objects clearly. Move to high ground (at least 10 feet or more above the desert floor) to get above

the superheated air close to the ground to overcome the mirage effect.
*Find shade! Get out of the sun!
*Place something between you and the hot ground.
*If water is scarce, do not eat.

Deserts are arid, barren regions of the earth. Successful desert operations require modifications to equipment and tactics as the environment can profoundly affect military operations. There are four main classifications of deserts: mountain deserts, rocky plateau deserts, polar deserts, and sandy or dune deserts. The common denominator of all four desert classifications is the lack of significant annual precipitation. Desert bedrock may have a layer of sand, ice, or gravel over it. Topsoil often has eroded or never formed due to any combination of a lack of water, heat, cold, or wind. Common land features in the desert consist of sand dunes, escarpments, dry riverbeds, and depressions. The lack of annual precipitation and topsoil leads to reduced wildlife and plant densities, often with unique and potentially hazardous adaptations, that give deserts their characteristic barren appearance.

Desert regions feature the greatest extremes of temperatures and weather volatility. Temperatures vary from extreme highs to extreme lows based upon the latitude and season. Temperatures can exceed 136 degrees Fahrenheit (57 degrees Celsius) in the deserts of Mexico and Libya to minus 128 degrees Fahrenheit (minus 88 degrees Celsius) in the Antarctic Polar desert. In some deserts, day-to-night temperature fluctuation exceeds 70 degrees Fahrenheit (21 degrees Celsius). Even the slightest rainfall in the desert can result in flash flooding due to lack of vegetation and the soil's inability to absorb the precipitation.

Key terrain in the desert is largely determined by the presence of water or terrain that restricts movement. The few roads available may become key terrain, especially when the desert floor cannot support wheeled vehicle traffic. Water sources are vital, especially if a force cannot sustain long-distance resupply. Where they exist, defiles play an important role. For example, in the western desert of Libya, an escarpment that parallels the coast forms a barrier to movement except through a few passes. Control of passes equates to control of key terrain. Similar escarpments exist in Saudi Arabia and Kuwait.

Mountain Deserts

Mountain deserts typically contain scattered ranges, areas of barren hills, or mountains, separated by dry flat basins. High ground may rise gradually or abruptly from flat areas, to a height of several thousand feet above sea level. Most of the infrequent rainfall occurs on high ground and runs off in the form of flash floods, eroding deep gullies and ravines and depositing sand and gravel around the edges of the basins. Water evaporates rapidly, leaving the land barren, although short-lived vegetation growth and flowering sometimes occurs. Shallow lakes may develop but will generally have high salt content due to high evaporation rates. The Great Salt Lake in Utah and the Dead Sea are examples of mountain desert lakes.

Rocky Plateau Deserts

Rocky plateau deserts are extensive flat areas with solid or broken rock at or near the surface. They can have wet or dry, steep-walled eroded valleys, dry riverbeds, gulches, or canyons. Narrow valleys can be extremely dangerous to personnel and equipment due to flash flooding after rains. The California National Training Center and the Syrian Golan Heights are both located in rocky plateau deserts.

Polar Deserts

The Antarctic and Arctic polar regions are the largest deserts on earth, each covering 5.5 million and 5.4 million square miles (8.8 million kilometers and 8.7 million kilometers), respectively. The Arctic region is located at the Arctic Circle (latitude 66° 32'N) and includes the northern continental fringes of North America, Iceland, coastal Greenland, and the Arctic coast of Eurasia. With long, severe winters and short, cool summers, the southern polar desert region encompasses the continent of Antarctica. In the Arctic Circle the sun never sets on the summer solstice and the sun never rises during the winter solstice (with the reverse occurring in Antarctica). The mean monthly temperature of the warmest month falls between 32 degrees and 50 degrees Fahrenheit (between 0 and 10 degrees Celsius). Annual precipitation is less than eight inches (20 centimeters), but low rates of evaporation make the climate humid. Vegetation consists of low-growing grasses, lichens, mosses, and brush with treeless plains. Soils are developed and have a permanently frozen sublayer (permafrost) that seasonally thaws at the surface. Surface and subsurface soil drains poorly and creates muddy summertime conditions. The Arctic predominately consists of coastal plains, low-interior and high-interior plains, and lesser areas of low and high-relief mountains. Most development and infrastructure of military interest centers around ports and areas with valuable natural resources. Some prominent terrain of polar deserts consists of icecaps, glaciers, and overflow ice.

Sandy or Dune Deserts
Sandy or dune deserts are extensive flat areas covered with sand or gravel, the product of ancient deposits or modern wind erosion. Flat is a relative term, as some areas may contain sand dunes that are over 984 feet (300 meters) high and 9–12 miles (16–19 kilometers) long. Trafficability on this type of terrain will depend on windward or leeward gradients of the dunes and the texture of the sand. Other areas, however, may be totally flat for distances of 2 miles (three kilometers) or more. Plant life may vary from none to scrub, reaching over 6.5 feet (two meters) high. Examples of this type of desert include the ergs of the Sahara, the Empty Quarter of the Arabian Desert, areas of California and New Mexico, and the Kalahari in South Africa.

Desert Weather

Lack of precipitation is the defining characteristic of the desert; desert environments receive less than 10 inches (25 centimeters) of sporadic rainfall annually. Desert rainfall varies from one day in the year to intermittent showers throughout the winter. Severe thunderstorms occur in the desert and bring heavy rain which can result in flash flooding. Rainstorms tend to be localized, affecting only a few square kilometers at a time.

The air stability in desert regions varies due to the extreme fluctuations between day and night temperatures. At night and early morning, the desert air is usually stable. High desert temperatures in the middle of the day decreases density and creates extremely unstable air. The three types of air stability are:

*Unstable (lapse). This condition exists when air temperature decreases with altitude. In the desert, this mostly occurs between late morning and early evening.
*Neutral. This condition exists when air temperature does not change with altitude. In the desert, this mostly occurs during early morning and early evening.
*Stable (inversion). This condition exists when the air temperature increases with altitude. In the desert, this mostly occurs between late evening and early morning.

High winds occur in certain desert seasons and can greatly affect operations due to reduced visibility. Blowing sand limits visibility. Reduced visibility can limit aviation support, inhibit the effectiveness of obscurants, cause health concerns, and increase the risk of accidents.

Desert Terrain

Desert terrain has many variations ranging from nearly flat, with high trafficability, to impassable mountain ranges. Units preparing to deploy to specific desert regions should seek detailed information on terrain prevalent in the expected area of operations.

Desert terrain can canalize operations due to poor cross-country mobility, limited hardened roads and trails, and the lack of bridges and hardened crossing points. Some terrains inhibit cross-country mobility with soft sand, rocky areas, fractured ice sheets, glaciers, salt flats, and marshy areas. Such terrains create poor trafficability. Roads in the desert are usually scarce, poorly maintained, and primitive. Limited infrastructure reduces hardened crossing points over or through water features.

The steep slopes of dunes and rock-strewn mountains can severely restrict vehicular movement. Dry riverbeds compartmentalize terrain. The banks of these stream beds can be steep and loose which severely limits operations. Slopes covered in rocks can hinder vehicles if rocks easily dislodge.

Vast, glaciated areas present additional hazards in the polar and Antarctic deserts. Such hazards can include hidden crevasses, fractured ice, and snow avalanches. Operations in these deserts require special training and equipment.

The lack of terrain features in many deserts can make land navigation extremely difficult. In deserts without significant terrain features, The Professional Citizen must use dead reckoning navigation skills and verify their position with GPS devices when available.

Key terrain in the desert can be any man-made feature, mountain pass, source of water, or high ground. Because there are few man-made features in the desert, those that do exist can become important, perhaps even key. Passes through steep terrain that enable vehicle passage often are key terrain due to their relative scarcity. The high ground in desert terrain is usually key terrain. The relative flatness and great distances of some deserts make even large sand dunes dominant features.

In desert regions, water sources, and the terrain that dominates access to water, are likely to be key terrain. Potable water sources, such as oases and wells, provide water resupply and tend to concentrate indigenous life. The logistics associated with water procurement and resupply can make a water source located in proximity to forces the key terrain for mission success.

The relatively few improved routes in many deserts may make them key terrain. These routes can become critical LOCs necessary for sustainment/logistics units to maintain tempo and operational reach.

Jungle Movement
When conducting movement in a jungle environment, The Professional Citizen must consider the following:

Jungle environment is the hardest to work in

*There is less likelihood of recovery from beneath a dense jungle canopy than in other survival situations. Movement will be required.
*Avoid saltwater swamps if you can. If there are water channels through it, use a raft to cross it.

*Do not concentrate on the pattern of bushes and trees to your immediate front. Focus on the jungle further out and find natural breaks in the foliage. Look through the jungle, not at it. Stop and stoop down occasionally to look along the jungle floor.
*Move through the jungle. Do not fight the jungle. Turn your shoulders, shift your hips, bend your body, and shorten or lengthen your stride as necessary to slide between the undergrowth.
*Do not grasp at brush, vines when moving, or climbing slopes; they may have irritating spines or sharp thorns.
*Protect yourself from insects with netting, clothing, etc.
*Promptly treat wounds and scratches to avoid dangerous infection.

The jungle environment includes densely forested areas, grasslands, cultivated areas, and swamps. Jungles are classified as primary or secondary jungles based on the terrain and vegetation.

PRIMARY JUNGLES
These are tropical forests. Depending on the type of trees growing in these forests, primary jungles are classified either as tropical rain forests or as deciduous forests.

Tropical Rain Forests. These consist mostly of large trees whose branches spread and lock together to form canopies. These canopies, which can exist at two or three different levels, may form as low as 10 meters from the ground. The canopies prevent sunlight from reaching the ground, causing a lack of undergrowth on the jungle floor. Extensive above-ground root systems and hanging vines are common. These conditions, combined with a wet and soggy surface, make vehicular traffic difficult. Foot movement is easier in tropical rain forests than in other types of jungle. Except where felled trees or construction make a gap in the canopy of the rain forest, observation from the air is nearly impossible. Ground observation is generally limited to about 50 meters (55 yards).

Deciduous Forests. These are found in semitropical zones where there are both wet and dry seasons. In the wet season, trees are fully leaved; in the dry season, much of the foliage dies. Trees are generally less dense in deciduous forests than in rain forests. This allows more rain and sunlight to filter to the ground, producing thick undergrowth. In the wet season, with the trees in full leaf, observation both from the air and on the ground is limited. Movement is more difficult than in the rain forest. In the dry season, however, both observation and trafficability improve.

SECONDARY JUNGLES

These are found at the edge of the rain forest and the deciduous forest, and in areas where jungles have been cleared and abandoned. Secondary jungles appear when the ground has been repeatedly exposed to sunlight. These areas are typically overgrown with weeds, grasses, thorns, ferns, canes, and shrubs. Foot movement is extremely slow and difficult. Vegetation may reach a height of 2 meters. This will limit observation to the front to only a few meters.

COMMON JUNGLE FEATURES

SWAMPS

These are common to all low jungle areas where there is water and poor drainage. There are two basic types of swamps: mangrove and palm.

Mangrove Swamps: These are found in coastal areas wherever tides influence water flow. The mangrove is a shrub-like tree which grows 1 to 5 meters high. These trees have tangled root systems, both above and below the water level, which restrict movement to foot or small boats. Observation in mangrove swamps, both on the ground and from the air, is poor. Concealment is excellent.

Palm Swamps: These exist in both salt and freshwater areas. Like movement in the mangrove swamps, movement through palm swamps is mostly restricted to foot (sometimes small boats). Vehicular traffic is nearly impossible except after extensive road construction by engineers. Observation and fields-of-fire are very limited. Concealment from both air and ground observation is excellent.

SAVANNA

This is a broad, open jungle grassland in which trees are scarce. The thick grass is broad-bladed and grows 1 to 5 meters high. Movement in the savanna is generally easier than in other types of jungle areas, especially for vehicles. The sharp-edged, dense grass and extreme heat make foot movement a slow and tiring process. Depending on the height of the grass, ground observation may vary from poor to good. Concealment from air observation is poor for both troops and vehicles.

BAMBOO

This grows in clumps of varying size in jungles throughout the tropics. Large stands of bamboo are excellent obstacles for wheeled or tracked vehicles. Troop movement through bamboo is slow, exhausting, and noisy. Troops should bypass bamboo stands if possible.

Most Americans, especially those raised in cities, are far removed from their pioneer ancestors, and have lost the knack of taking care of themselves under all conditions. It would be foolish to say that, without proper training, they would be in no danger if lost in the jungles of Southeast Asia, South America, or some Pacific Island. On the other hand, they would be in just as much danger if lost in the mountains of western Pennsylvania or in other undeveloped regions of our own country. The only difference would be that a man is less likely to panic when he is lost in his homeland than when he is lost abroad.

JUNGLE HAZARDS

Effects of Climate
The discomforts of tropical climates are often exaggerated, but it is true that the heat is more persistent. In regions where the air contains a lot of moisture, the effect of the heat may seem worse than the same temperature in a dry climate. Many people experienced in jungle living feel that the heat and discomfort in some US cities in the Summertime are worse than the climate in the jungle.

Strange as it is, there may be more suffering from cold in the tropics than from the heat. Of course, very low temps do not occur, but chilly days and nights are common. In some jungles, in winter months, the nights are cold enough to require a wool blanket or poncho liner for sleeping.

Rainfall in many parts of the tropics is much greater than that in most areas of the temperate zones. Tropical downpours usually are followed by clear skies, and in most places the rains are predictable at certain times of the day. Except in those areas where rainfall may be continuous during the rainy season, there are not many days when the sun does not shine part of the time.
People who live in the tropics usually plan their activities so that they are able to stay under shelter during the rainy and hotter portions of the day. After becoming used to it, most tropical dwellers prefer the constant climate of the torrid zones to the frequent weather changes in colder climates.

Insects
Malaria-carrying mosquitoes are probably the most harmful of the tropical insects. Folks can contract malaria if proper precautions are not taken.
Mosquitoes are most prevalent early at night and just before dawn. Folks must be especially cautious at these times. Malaria is more common in populated areas than in

uninhabited jungle, so soldiers must also be especially cautious when operating around villages. Mud packs applied to mosquito bites offer some relief from itching.

Wasps and bees may be common in some places, but they will rarely attack unless their nests are disturbed. When a nest is disturbed, the troops must leave the area and reassemble at the last rally point. In case of stings, mud packs are helpful. In some areas, there are tiny bees, called sweat bees, which may collect on exposed parts of the body during dry weather, especially if the body is sweating freely. They are annoying but stingless and will leave when sweating has completely stopped, or they may be scraped off with the hand.

The larger centipedes and scorpions can inflict stings which are painful but not fatal. They like dark places, so it is always advisable to shake out blankets before sleeping at night, and to make sure before dressing that they are not hidden in clothing or shoes. Spiders are commonly found in the jungle. Their bites may be painful but are rarely serious. Ants can be dangerous to injured folks lying on the ground and unable to move. Wounded individuals should be placed in an area free of ants.

In Southeast Asian jungles, the rice borer moth of the lowlands collects around lights in great numbers during certain seasons. It is a small, plain-colored moth with a pair of tiny black spots on the wings. It should never be brushed off roughly, as the small, barbed hairs of its body may be ground into the skin. This causes a sore, much like a burn, that often takes weeks to heal.

Leeches
Leeches are common in many jungle areas, particularly throughout most of the Southwest Pacific, Southeast Asia, and the Malay Peninsula. They are found in swampy areas, streams, and moist jungle country. They are not poisonous,

but their bites may become infected if not cared for properly. The small wound that they cause may provide a point of entry for the germs which cause tropical ulcers or "jungle sores." Folks working in the jungle should watch for leeches on the body and brush them off before they have had time to bite. When they have taken hold, they should not be pulled off forcibly because part of the leech may remain in the skin. Leeches will release themselves if touched with insect repellent, a moist piece of tobacco, the burning end of a cigarette, a coal from a fire, or a few drops of alcohol.

Straps wrapped around the lower part of the legs ("leech straps") will prevent leeches from crawling up the legs and into the crotch area. Trousers should be securely tucked into the boots.

Snakes
A person in the jungle probably will see very few snakes. When he does see one, the snake most likely will be making every effort to escape.

If folks should accidentally step on a snake or otherwise disturb a snake, it will probably attempt to bite. The chances of this happening to soldiers traveling along trails or waterways are remote if soldiers are alert and careful. Most jungle areas pose less of a snakebite danger than do the uninhabited areas of New Mexico, Florida, or Texas. This does not mean that soldiers should be careless about the possibility of snakebites, but ordinary precautions against them are enough. Folks should be particularly careful when clearing ground.

Treat all snakebites as poisonous.

Crocodiles and Caymans
Crocodiles and Caymans are meat-eating share of crocodiles, but there are few reptiles which live in tropical areas. Authenticated cases of crocodiles actually

"Crocodile-infested rivers and swamps" is a catch-phrase often found in stories about the tropics. Asian jungles certainly have their attacking humans. Caymans, found in South and Central America, are not likely to attack unless provoked.

Poisonous Vegetation

Another area of danger is that of poisonous plants and trees. For example, nettles, particularly tree nettles, are one of the dangerous items of vegetation. These nettles have a severe stinging that will quickly educate the victim to recognize the plant. There are ringas trees in Malaysia which affect some people in much the same way as poison oak. The poison ivy and poison sumac of the continental US can cause many of the same type troubles that may be experienced in the jungle. The danger from poisonous plants in the woods of the US eastern seaboard is similar to that of the tropics. Thorny thickets, such as rattan, should be avoided as one would avoid a blackberry patch. Some of the dangers associated with poisonous vegetation can be avoided by keeping sleeves down and wearing gloves when practical.

HEALTH AND HYGIENE

The climate in tropical areas and the absence of sanitation facilities increase the chance that folks may contract certain diseases. Prevention of disease is fought with good sanitation practices and preventive medicine. In past times of war and natural disasters, diseases accounted for a significantly high percentage of casualties.

Waterborne Diseases

Water is vital in the jungle and is usually easy to find. However, water from natural sources should be considered contaminated. Water purification procedures must be Taught and practiced by everyone. (Water treatment is covered in Chapter 1.) Germs of serious diseases, like dysentery, are found in impure water. Other waterborne diseases, such as blood fluke, are caused by exposure of an open sore to impure water.

Fungus Diseases
These diseases and conditions are caused by poor personal health practices. The jungle environment promotes fungus and bacterial diseases of the skin and warm water immersion skin diseases. Bacteria and fungi are tiny plants which multiply fast under the hot, moist conditions of the jungle. Sweat soaked skin invites fungus attack. The following are common skin diseases that are caused by long periods of wetness of the skin.

Warm Water Immersion Foot. This condition occurs usually where there are many creeks, streams, and canals to cross, with dry ground in between. The bottoms of the feet become white, wrinkled, and tender. Walking becomes painful.

Chafing. This condition occurs when folks must often wade through water up to their waists, and the trousers stay wet for hours. The crotch area becomes red and painful to even the lightest touch. If you see someone in your group walking like a penguin, chances are they have crotch or inner thigh chaffing.

Most skin conditions are easily treated by letting the skin dry.

HEAT INJURIES
These result from high temperatures, high humidity, lack of air circulation, and physical exertion. All folks should be trained to prevent heat disorders.

WATER CROSSINGS

Almost every description can be applied to rivers and streams. They may be shallow or deep, slow or fast moving, narrow or wide. Before crossing a river or stream, a good plan will need to be developed. The first step is to look for a high place from which to get a good view of the river or stream. From here, you can look for a place to cross.

Rivers and Streams

Good crossing locations include:

*A level stretch where it breaks into several channels. Two or three narrow channels are usually easier to cross than a wide river.
*A shallow bank or sandbar. If possible, select a point upstream from the bank or sandbar so that the current will carry them to it if they lose their footing.
*A course across the river that leads downstream so that isolated persons will cross the current at about a 45- degree angle.

The following areas possess potential hazards; avoid them if possible:

*A ledge of rocks that crosses the river. This often indicates dangerous rapids or canyons.
*An estuary of a river, because it is normally wide, has strong currents and is subject to tides. These tides can influence some rivers many miles from their mouths; personnel should go back upstream to an easier crossing site.
*Eddies, which can produce a powerful backward pull downstream of the obstruction causing the eddy and pull personnel under the surface.

Rapids

To swim across a deep, swift river, swim with the current; do not swim against the current. Try to keep the body horizontal to the water. This will reduce the danger of being pulled under. In fast, shallow rapids, lie on your back, feet pointing downstream, finning your hands alongside your hips. This action will increase buoyancy and help steer away from obstacles. Keep your feet up to avoid getting them caught by rocks.

In deep rapids, lie on your stomach, head downstream, and angling toward the shore. Watch for obstacles and be careful of backwater eddies and converging currents, as they often contain dangerous swirls. Converging currents occur where new watercourses enter the river or where water has been diverted around large obstacles such as small islands. To ford a swift, treacherous stream, apply the following steps:

*Step 1. Find a strong pole about 3 inches in diameter and 7 to 8 feet long to help ford the stream.
*Step 2. Grasp the pole and plant it firmly on the upstream side to break the current. Plant feet firmly with each step, and move the pole forward a little downstream from its previous position, but still upstream.
*Step 3. With the next step, place foot below the pole. Keep the pole well slanted so that the force of the current keeps the pole against their shoulder. Cross the stream in such a manner that the downstream current is being crossed at a 45-degree angle.

Example of crossing a stream with the aid of a pole

If there are other people with the isolated person, ensure that they all cross the stream together. Position the heaviest person on the downstream end of the pole and the lightest person on the upstream end. In using this method, the upstream person breaks the current, and those below can move with relative ease in the eddy formed by the upstream person. If the upstream person temporarily loses footing, the others can hold steady while the upstream person regains footing.

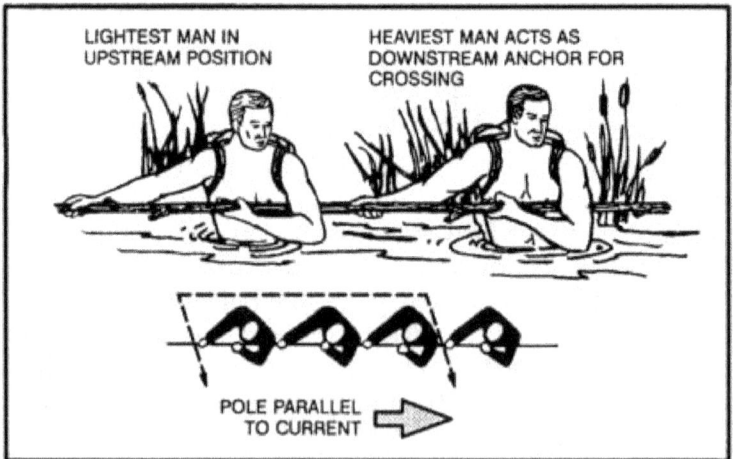

LIGHTEST MAN IN
UPSTREAM POSITION

HEAVIEST MAN ACTS AS
DOWNSTREAM ANCHOR FOR
CROSSING

POLE PARALLEL
TO CURRENT

Utilizing a pole for multiple people crossing

If there are three or more people crossing the stream and a rope is available, the technique shown below can be used. The length of the rope must be three times the width of the stream.

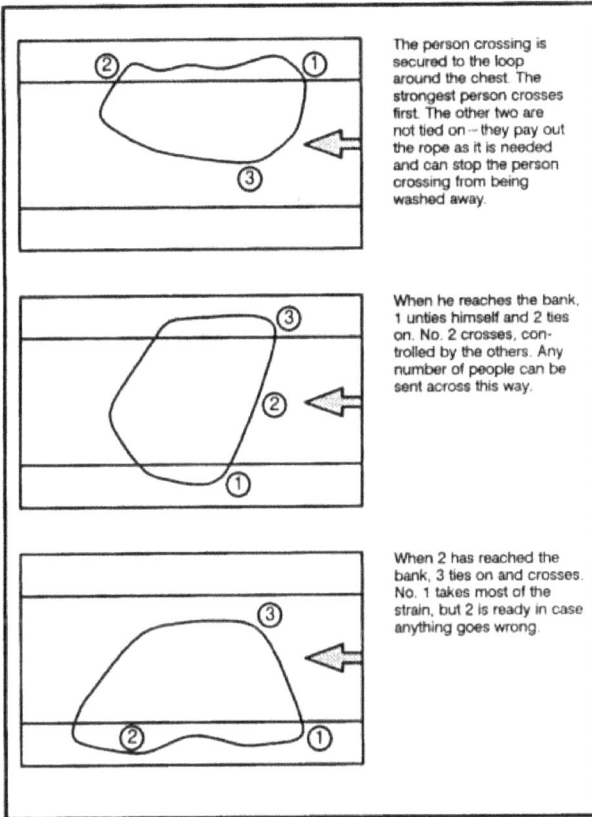

The person crossing is secured to the loop around the chest. The strongest person crosses first. The other two are not tied on – they pay out the rope as it is needed and can stop the person crossing from being washed away.

When he reaches the bank, 1 unties himself and 2 ties on. No. 2 crosses, controlled by the others. Any number of people can be sent across this way.

When 2 has reached the bank, 3 ties on and crosses. No. 1 takes most of the strain, but 2 is ready in case anything goes wrong.

River crossing using rope

RAFTS

Rafts are going to be a useful piece of equipment (along with the knowledge to make one) that enables a safer water crossing when involved in a survival scenario to cross a large body of water, with equipment, or if injured.

If two ponchos are available, construct a brush raft or an Australian poncho raft. Using either of these rafts, equipment can be safely floated across a slow-moving stream or river. Let's look at a couple examples of building field expedient rafts.

Bush Raft
The bush raft is constructed out of ponchos, fresh green brush, two small saplings, and rope or vine.
as follows:

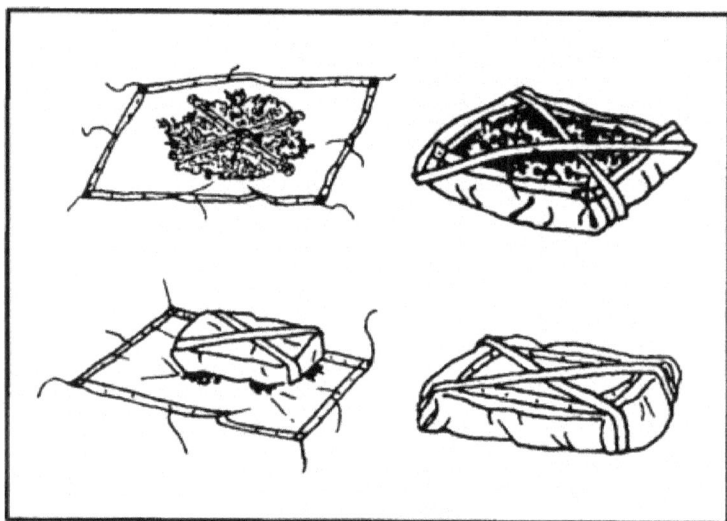

Example of a Bush Raft

*Step 1. Push the hood of each poncho to the inner side and tightly tie off the necks using the drawstrings.
*Step 2. Attach the ropes or vines at the corner and side grommets of each poncho. Make sure they are long enough to cross to and tie with the others attached at the opposite corner or side.
*Step 3. Spread one poncho on the ground with the inner side up. Pile fresh, green brush (no thick branches) on the

poncho until the brush stack is about 18 inches high. Pull the drawstring up through the center of the brush stack.
*Step 4. Make an X-frame from two small saplings and place it on top of the brush stack. Tie the X-frame securely in place with the poncho drawstring.
*Step 5. Pile another 18 inches of brush on top of the X-frame, and then compress the brush slightly.
*Step 6. Pull the poncho sides up around the brush and, using the ropes or vines attached to the corner or side grommets, tie them diagonally from corner to corner and from side to side.
*Step 7. Spread the second poncho, inner side up, next to the brush bundle.
*Step 8. Roll the brush bundle onto the second poncho so that the tied side is down. Tie the second poncho around the brush bundle in the same manner as the first poncho was tied around the brush.
*Step 9. Place the raft in the water with the tied side of the second poncho facing up.

Australian Poncho Raft
The Australian poncho raft is constructed when there is no time to gather brush for a brush raft. This raft, although more waterproof than the poncho bush raft, will only float about 77 pounds of equipment. To construct this raft, use two ponchos, two rucksacks, two 4-foot poles or branches, and ropes, vines, bootlaces, or comparable material as follows:

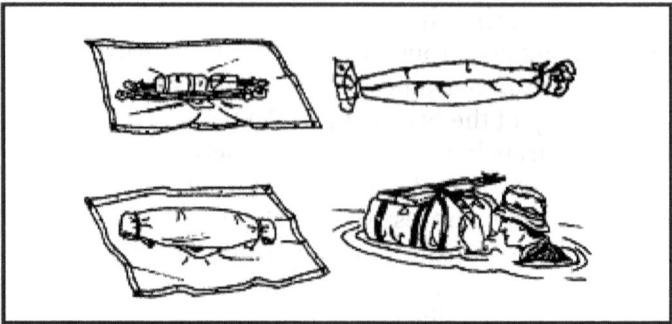

Example of the Australian Poncho Raft

*Step 1. Push the hood of each poncho to the inner side and tightly tie off the necks using the drawstrings.

*Step 2. Spread one poncho on the ground with the inner side up. Place and center the two 4-foot poles on the poncho about 18 inches apart.

*Step 3. Place rucksacks, packs, or other equipment between the poles. Also, place other items that need to be kept dry between the poles. Snap the poncho sides together.

*Step 4. Use a friend's help to complete the raft. Hold the snapped portion of the poncho in the air and roll it tightly down to the equipment. Make sure to roll the full width of the poncho.

*Step 5. Twist the ends of the roll to form pigtails in opposite directions. Fold the pigtails over the bundle and tie them securely in place using ropes, bootlaces, or vines.

*Step 6. Spread the second poncho on the ground, inner side up. If more buoyancy is needed, place some fresh green brush on this poncho.

*Step 7. Place the equipment bundle, tied side down, on the center of the second poncho. Wrap the second poncho around the equipment bundle following the same procedure as used for wrapping the equipment in the first poncho.

*Step 8. Tie ropes, bootlaces, vines, or other binding material around the raft about 12 inches from the end of each pigtail. Place and secure weapons on top of the raft.

*Step 9. Tie one end of a rope to an empty canteen and the other end to the raft. This will help to tow the raft.

So, I think you can see that the materials used for creating a raft come right out of the ten essentials list from Chapter 1. There are so many reasons why that list was created and why it is so vital to your overall preparedness for survival situations. There is no excuse to not have some form of those ten essentials on you. Along with this is the need to practice. Putting in the effort now, and experiencing ahead of time, the discomfort of these situations will pay dividends down the road. Do the work now, you'll be thankful that you did.

Navigation:

Where are we? How do we find our way from where we are to where we want to get to? How far are we from camp? These are some of the more common questions asked when dealing with navigation in the backcountry.

Tools of the trade to get started.

I know that for many, map and compass work is a foreign subject. Especially in today's electronic age where everyone has a GPS in their car, watch, and phone. BUT, what happens when that battery dies, or you drop and break it, or you are under heavy tree cover or in a canyon? Do you know the basics of navigation? We will cover some basics here that will allow you to get back on track and find your way again in a survival situation.

First, let's cover a few quick definitions:

-Orientation: This is the act of determining your location on earth. It requires good map and compass skills along with being able to use an altimeter and GPS. Folks who generally spend a good amount of time in the backcountry usually gain these skills naturally.

-Navigation: This would be the act of determining the location of your objective and keeping yourself going in the right direction. Like orientation, navigation requires good map and compass skills and is a necessity for backcountry travel.

-Routefinding: This is the art of being able to select and follow the most appropriate path for not just your ability level, but also the path appropriate for the ability of the group and for the gear available to you. Being a good route finder takes a lot of hard work. You need good judgement, experience, and a bit of instinct.

Basic Navigation Tools

-Maps
-Compass
-Pencil/pens
-Map Markers
-Map Case
-Pace Beads
-Notebook
-Protractor
-Straight Edge

Preparing for your Outing

So, you can see how all three of these are important to fully understanding the art of land navigation, especially in a survival situation. But it all starts before you even head out your door. You should be gathering information about the area including knowing what the surrounding are looks like, approximate times to known points, elevation gain, and more.

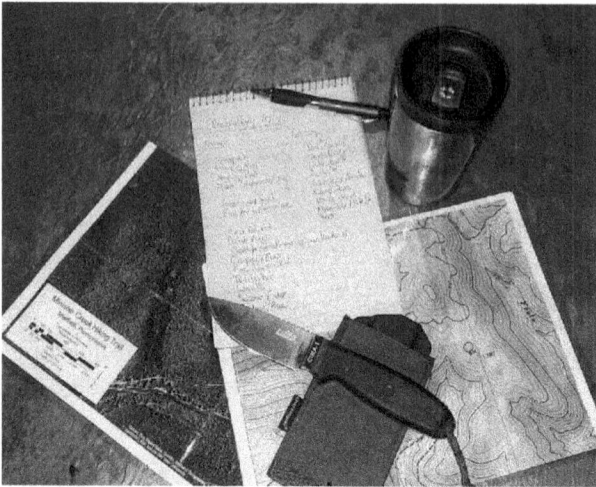
Routefinding begins at home.

If your information comes from a guidebook or another person's notes, then make sure you are notating everything on your maps and in a notebook for reference. Note things like trail junctions, river crossings, vantage points, cliffs, dwellings, and more.

Before you head out you should have a mental image of the route. From experience, and from all the resources you used about the route, you'll have a better understanding of how to make the terrain work to your advantage. Most times the route going out is the same as that you are returning on, but if it is different, then you need to be sure that you have researched that route too.

Maps
Your map is a phenomenal piece of information presented in a way that is both easy to understand and easy to carry. No one should travel without being able to understand one. Always be sure to take note of the publication dates on your map because trails, roads, and many other features may have changed over time. Here is a list of the different types of maps.

-Relief maps
-Land Management/Recreation Maps
-Sketch Maps
-Guidebook maps
-Topographic maps

So, let's talk about topographic maps. A topographic map provides information on the existence, the location of, and the distance between ground features, such as populated places and routes of travel and communication. It also indicates variations in terrain, heights of natural features, and the extent of vegetation coverage. Topographic maps portray terrain features (through the use of contour lines) as well as the horizontal positions of the features being represented. The vertical positions, or relief, are represented by contour lines on topographic maps. On maps showing relief, the elevations and contours are measured from a specific vertical datum plane, usually meaning sea level.

At the heart of a topographic map is its contour lines. Each one indicates a constant elevation change as it follows the landscape. A map's contour interval is the difference in elevation between two adjacent contour lines. So, one of the most important pieces of information derived from this is whether you will be traveling uphill or downhill. If your route crosses lines of increasingly higher elevation, then you are going uphill. If your route crosses lines of decreasing elevation, then you are going downhill. Flat or sidehill travel is indicated by a route that crosses no lines, remaining within a single contour interval.

But this is only the start of the picture that contour lines show us. They also show cliffs, mountain passes, ravines, summits, and more. You will get better and better at interpreting these lines when comparing them to the actual terrain as it is represented on your map. The idea here is to be able to glance at a map and have a pretty good mental image as to just what the terrain actually looks like.

Terrain Features

The topographical map is an accurate representation of the ground, we just have to know how to "read" the map. Learning to read a map is kind of like learning a new language, but with practice it becomes second nature and soon the map features will make sense and you'll be able to pull it all together.

Let's cover some major terrain features:

***Flat areas:** No contour lines at all
***Gentle Slops:** Widely spaced contour lines
***Steep slope:** Closely spaced contour lines
***Cliffs:** Contour lines extremely close together or touching
***Valleys, ravines, gullies, and couloirs:** Contour lines in a pattern of U's for gentle, rounded valleys or gullies; V's for sharp valleys or gullies. The U's or V's point uphill, in the direction of higher elevation.
***Ridge or spur:** Contour lines in a pattern of U's for gentle, rounded ridges; V's for sharp ridges. The U's or V's point downhill, in the direction of lower elevation.
***Summit or Peak:** A concentric pattern of contour lines, with the summit being the innermost and highest ring. Peaks often are also indicated by X's, elevations, benchmarks (BM's), or a triangle symbol.
***Cirques or bowls:** Patterns of contour lines forming a semicircle (or as much as a three-quarters circle), rising from a low spot in the center of the partial circle to form a natural amphitheater at the head of a valley.
***Saddle, col, or pass:** An hourglass shape, with higher contour lines on each side, indicating a low point on a ridge.

The margin on a topographical map holds important information such as the date of publication, name, contour interval, map scale, North and magnetic declination.

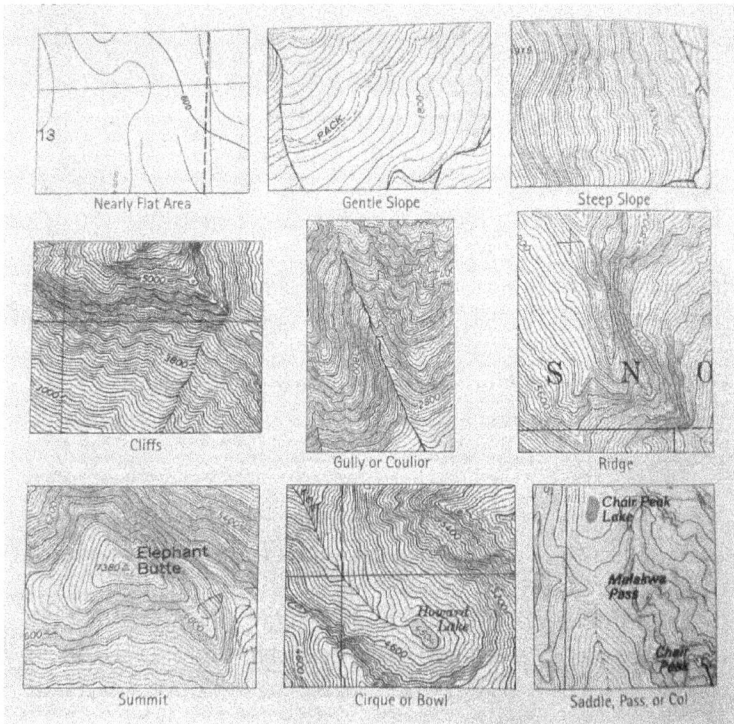

Basic terrain features.

There are some cautionary thoughts to keep in mind as you look at your topographical maps since they do have some limitations. Your map will not show all the terrain features just due to the simple fact that there is only so much room available to put information. An example would be smaller cliffs that, if shorter than the contour line intervals, may come as a surprise to you on the trail. Also check the dates on your map, if it is an older date, then be wary of human intervention such as logging roads and power line cuts. You may need to supplement these older maps with information taken from online sources, current aerial imagery, guidebooks, or even other maps. If you do find differences, then definitely notate them.

Navigation
Generally, there are three primary means of navigation available to: dead reckoning, terrain association, and map/compass.

*__Dead reckoning__ consists of two fundamental steps. The first step is to use a protractor and graphic scales to determine the direction and distance from one point to another on a map. The second step is to use a compass and some means of measuring distance to apply this information on the ground. In other words, it begins with the determination of a polar coordinate on a map and ends with the act of finding it on the ground.

Dead reckoning along a given route is the application of the same process used by a mapmaker to establish a measured line of reference with which to construct the framework of the map. Resection or intersection can be easily undertaken at any time to either determine, or confirm, precise locations along or near their route. Between these position fixes, folks can establish their location by measuring or estimating the distance traveled along the azimuth being followed from the previous known point. They might use pacing, a vehicle odometer, or the application of elapsed time for this purpose, depending upon the situation.

Most dead reckoned movements do not consist of single straight-line distances because the tactical and navigational aspects of the terrain, the enemy situation, natural and manmade obstacles, time, and safety factors cannot be ignored. Another reason most dead reckoning movements are not single straight-line distances is because compasses and pace counts are imprecise measures. Error from them compounds over distance; therefore, isolated persons could soon be far from their intended route even if they performed the procedures correctly. The only way to counteract this phenomenon is to reconfirm the location by terrain association or resection. Routes planned for dead reckoning generally consist of a series of straight-line distances between several checkpoints.

There are two advantages to dead reckoning. First, dead reckoning is easy to teach and learn. Second, it can be a highly accurate way of moving from one point to another if done carefully over short distances, even where few external cues are present to guide movements. I have had great success with this method of navigation. By breaking down my movement into small sections, I have been able to accurately and successfully navigate large swaths of wilderness areas simply by taking my time, paying attention, and referencing my maps many times in each section.

During daylight, across open country, along a specified magnetic azimuth, folks should never walk with the compass in the open position and in front of them. Because the compass will not stay steady or level, it does not give an accurate reading when held or used this way. Begin at the start point and face with the compass in the proper direction, then sight in on a landmark located on the correct azimuth to be followed. Close the compass and proceed to that landmark. Repeat the process as many times as necessary to complete the straight-line segment of the route. Even if you do it every 20 yards in heavily forested areas, this system of map orientation and azimuth taking works every time.

The landmarks selected for these purposes are called steering marks, and their selection is crucial to success in dead reckoning. Steering marks should never be determined from a map study. They are selected as the march progresses and are commonly on or near the highest points visible along the azimuth line being followed. They may be uniquely shaped trees, rocks, hilltops, posts, towers, and buildings—anything that can be easily identified. If folks do not see a good steering mark to the front, they might use a back azimuth to some feature behind them until a good steering mark appears out in front. Characteristics of a good steering mark include the following:

*It must have some characteristics, such as color, shade of color, size, or shape (preferably all four), that will assure folks that it will continue to be recognized as they approach it.
*It is the most distant object available along the line of the march. This enables isolated persons to move farther with fewer references to the compass. If there are many options, they should select the highest object. A higher mark is not as easily lost to sight as is a lower mark that blends into the background as it is approached. A steering mark should be continuously visible as folks move toward it.
*A steering mark selected at night must have even more unique shapes than one selected during daylight. As darkness approaches, colors disappear, and objects appear as black or gray silhouettes. Instead of seeing shapes, only the general outlines are visible that may appear to change as they move, and objects must be seen from slightly different angles.

Dead reckoning without natural steering marks is used when the travel area is devoid of features, or when visibility is poor. At night, it may be necessary to send a member of the unit out in front of the unit's position to create steering marks in order to proceed. The position should be as far out as possible to reduce the number of chances for error during movement. Arm-and-hand signals or a radio may be used in placing the isolated person on the correct azimuth. After being properly located, move forward to this position and repeat the process until some steering marks can be identified or until the objective is reached.

When handling obstacles and detours on the route, follow these guidelines:

*When an obstacle forces folks to leave their original line of march and take up a parallel one, always return to the original line as soon as the terrain or situation permits.
*To turn clockwise (right) 90°; 90° must be added to your original azimuth. To turn counterclockwise (left) 90° from their current direction, 90° must be subtracted from your present azimuth.
*When making a detour, be certain that only paces taken toward the destination are counted as part of the forward progress. The forward progress paces should not be confused with the local pacing that takes place perpendicular to the route to avoid the problem area and in returning to the original line of travel after the obstacle has been passed.

Highly accurate distance estimates and precision compass work may not be required for a deliberate offset technique if the destination or an intermediate checkpoint is located on or near a large linear feature that runs nearly perpendicular to the direction of travel. Examples include roads or highways, railroads, power transmission lines, ridges, or streams. In these cases, folks should apply a deliberate error (offset) of about 10° to the azimuth they planned to follow and then move, using the lensatic compass as a guide, in that direction until they encounter the linear feature. They will know exactly which way to turn (left or right) to find their destination or checkpoint,

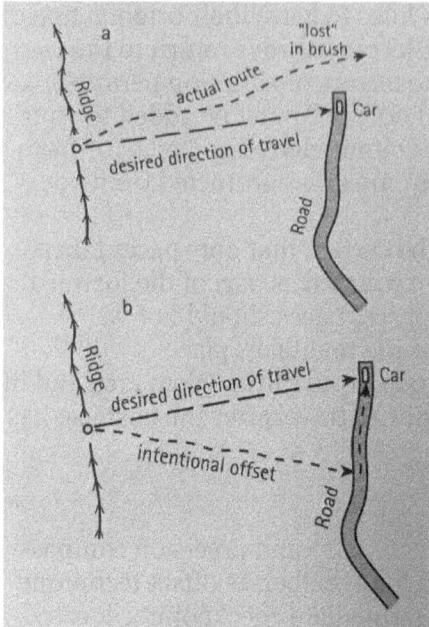

depending upon which way they planned their deliberate offset.

Because no one can move along a given azimuth with absolute precision, it is better to plan a few extra steps than to begin an aimless search for the objective once reaching the linear feature. This method also copes with minor compass errors and the slight variations that always occur in the earth's magnetic field.

There are disadvantages to dead reckoning. The farther the movement by dead reckoning without confirming the position in relation to the terrain and other features, the more errors that will accumulate in the movements. Therefore, folks should confirm and correct their estimated position whenever they encounter a known feature on the ground that is also on the map. Periodically, complete a resection using two or more known points to pinpoint and correct their position on the map. Pace counts or any type of distance, measurement should begin anew each time their position is confirmed on the map.

*Terrain Association: The technique of moving by terrain association is more forgiving of mistakes and far less time consuming than dead reckoning. It best suits those situations that call for movement from one area to another. Once an error has been made in dead reckoning, the folks will be off track. However, errors made using terrain association are easily corrected, because folks are comparing what they expected to see from the map to what they do see on the ground. Errors are anticipated and will not go unchecked. Folks can easily make adjustments based upon what they encounter. Periodic position fixing through either plotted or estimated resection will also make it possible to correct movements.

Identifying and Locating Selected Features
*Being able to identify and locate selected features, both on the map and on the ground, are essential to the success in moving by terrain association. The following rules may prove helpful:
*Be certain the map is properly oriented when moving along the route and use the terrain and other features as guides. The orientation of the map must match the terrain or it can cause confusion.
*When locating and identifying features to be used to guide movement, look for the steepness and shape of slopes, the relative elevations of various features, and the directional orientations in relation to the isolated person's position and to the position of other features that can be seen.
*Make use of the additional cues provided by water features and vegetation.

Using Handrails, Catching Features, and Navigational Attack Points
The following paragraphs discuss how to use handrails, catching features, and attack points to determine direction.

Handrails are linear features, such as roads or highways, railroads, power transmission lines, ridgelines, or streams that run roughly parallel to the isolated person's direction of movement. Instead of using precision compass work, isolated persons can rough compass (using the linear feature to follow the general compass direction) without the use of steering marks for as long as the feature travels with them on their right or left. It acts as a handrail to guide the way.

Catching features are when folks reach the point where either their route or the handrail changes direction, they must be aware that it is time to go their separate ways. Some prominent feature located near this point is selected to provide this warning. This is called a catching feature; it can also be used to tell when they have gone too far, acting as a backstop.

Navigational attack points These catching features may also be folks navigational attack point; this point is the place where area navigation ends and point navigation begins. From this last easily identified checkpoint, move cautiously and precisely along a given azimuth for a specified distance to locate the final objective. The selection of this navigational attack point is important. A distance of about 500 meters or less is most desirable.

Combining Techniques are often times the most successful way to navigate. Constant orientation of the map and continuous observation of the terrain in conjunction with compass-read azimuths, and distance traveled on the ground compared with map distance, when used together, make reaching a destination more certain. Folks should not depend entirely on compass navigation or map navigation.

*Using a Compass: A compass is a relatively simple device that can reveal at anytime and anyplace exactly the direction that you are heading. On a simple field exercise or outing you may never even need to look at your compass. But as the terrain and route becomes more complex the more you may need to use it.

Examples of different compasses.

The basic features of a compass include:

-A rotating magnetic needle, one end is different from the other so you can differentiate which end points north.
-A circular, rotating housing for the needle, filled with fluid that reduces the vibrations of the needle, making it more accurate.
-A dial around the housing with degrees marked on it from 0 to 360.
-An orienting arrow and a set of parallel meridian lines.
-An index line.
-A see through baseplate that includes a direction of travel line to point towards your objective.

index line

rotating housing with dial

orienting arrow

direction of travel line

base plate

magnetic needle

meridian lines

a

sighting notch

sighting mirror

declination arrow

lanyard

clinometer

ruler

b

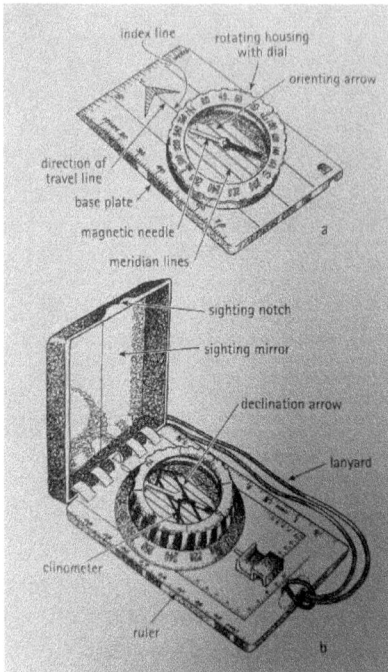

Then you may have some additional features including:

-An adjustable declination arrow.
-A sighting mirror.
-A ruler for measuring distances on your map.
A clinometer for use in measuring slope angle. This can be handy for determining the steepness of a slope for assessing avalanche terrain.
-A magnifying glass which can help with reading closely spaced contour lines.

For pure navigation work nothing beats a lensatic style compass like the Military Cammenga Lensatic Compass or the Silva Ranger Compass. These styles are durable, easy to use, accurate, and designed to work with a multitude of maps.

A point on compasses that is important to make. As we mentioned previously, there are many ways to navigate and many times just simply being able to find north and orient your map for it will be crucial in getting out of a survival situation. Keeping a small button style compass dummy corded in a BDU or Smock pocket or in an E&E kit can pay dividends in a real-life scenario.

Maps are drawn to a specific representation of a piece of the earth's surface. This representation is oriented or pointed to true north or in the direction of the North Pole located along the earth's rotational axis. The compass points to magnetic north or the north magnetic pole.

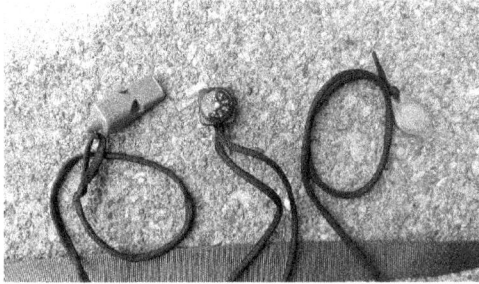

This is in the general direction of true north, but depending on where the map is drawn to on the earth's surface there is a variation between the two. This is an easterly or westerly variation and will be represented in different degrees. The declination diagram in the map's information section will provide the angular relationship between true north and magnetic north. This allows for making an adjustment to the map that is known as orienting the map.) The best way to orient a map is with a compass.

The lensatic compass has a needle with a north direction marked on the bottom inside of the compass. A button or wrist compasses may have floating dials or floating needles. To determine the heading, line up the north-seeking arrow over 360° by rotating the compass, then read the desired heading. Orienting a map with a floating needle compass is similar to the method used with the floating dial. The only exception is with the adjustment for magnetic variation. If magnetic variation is to the east, turn the map and the compass to the left (the north axis of the compass should be aligned with the map north) so that the magnetic north seeking arrow is pointing at the number of degrees on the compass which correspond with the angle of declination.

Use caution to ensure nothing (metal, mining, and ore) are in the area, the metal can alter your compass reading. I was once sitting on a wooden bridge with my map laid out orienting it with my compass and the needle kept bouncing around. I was confused at first but then realized that my compass was sitting on top of a nail head that was throwing it off. I've even had this happen next to a rifle, wristwatch, and along railroad tracks.

*Map Orientation: During a field exercise or outing it helps to hold your map open and orient it so that North on the map is actually lined up in the direction of true north. Do this is referred to as "orienting the map" and it is a good way to gain a better feel of where you are between the map and the countryside.

It is really a simple process where you set your compass to 0 or 360 degrees at the index line of your compass on your map, near its lowest left corner. Put the edge of your baseplate along the left edge of the map, with the direction of travel arrow (or line) pointing towards north on the map. Then you simply turn your map and compass together until the north seeking end of your compass needle is lined up with the pointed end of the declination arrow of the compass. You now have a properly oriented map and better view of the scene around you.

NAVIGATION TIP
Prior to navigation, it is always recommended to reorient your map. A map is oriented when it is in a horizontal position with its north and south corresponding to the north and south on the ground.

What if You Get Lost

I'll admit it, I have been lost twice, in-fact I found myself alone on solo backcountry trips and in a position where I had to do full on stop and assessments of the situation. One of these times involved having to overnight and wait till sun up to get back on track.

So, here is a brief story…
I was 19 years old on a solo traverse of the Uinta Mountain Range via the Highline Trail in NE Utah. About 30 miles in I was getting ready to find a spot to camp for the night and when looking at my map versus my surroundings I realized that I wasn't near the lake at the base of Leidy peak. I was near a lake and at the base of a mountain, but it wasn't the right one. After a couple of hours, I verified via terrain recognition (remember that from earlier) that I had ended up at the base of Marsh Peak about 4 miles south of where I was supposed to be. The mistake I made was I took the trail for granted and quit periodically looking at my map. I accidentally caught a mountaineer's trail and just kept going. While not a big mistake, it set me back almost a full day on the 102-mile trip. Thankfully I was prepared with extra rations and in a fairly water rich environment where my mistake didn't have negative consequences.

Anyways, folks get lost for various reasons. Some because they take the trail for granted and don't bring a map or compass with them, others fail to check on local conditions (recent blowdowns, avalanches, flooding, etc…) before setting off, some just don't pay attention to the route on the way in and can't find it on their way out. Some rely on the skill of their partner who is in the process of getting them lost.

Good navigators are never truly lost, maybe a might bit bewildered, but not really lost. And those good navigators have the good sense about them to always be sure to have enough food, water, clothing, and related gear to get them through a confusing spot.

So, what if you do get lost? The first rule is to STOP. Do not just forge ahead hoping for the best. You need to take the time to figure out where you are. Where was the last spot that you knew where you were. Was it just a few minutes ago or hours ago. You might be able to get back to that location with some intelligent guesses, moving cautiously, and paying attention to familiar landmarks.

Rarely do groups get really lost, the real danger comes to the person who is separated from their group. This is a major reason as to why you have not just a point man on the trail, but a designated rear guard as well keeping track of any stragglers.

Now, what should you do if you do find yourself lost? As stated above, you STOP. Take a look around looking for other members of your group, shout (a whistle is handy in this situation, you do have a whistle in that signaling kit, right...), and wait for a response. If there us only silence, then it's time to sit down, collect your thoughts, and put things into perspective.

Once you have collected yourself, start doing the right things. Look at your map, orient it, and try to determine your location, and start planning a route out in case you can't reconnect with your group. Mark your location with a rock cairn or other objects and start scouting out in all directions each time returning to the location you marked. Before it gets dark start gathering firewood, find water, and select a spot for spending the night. Not only does this keep you busy but it also helps in getting you settled in for the night.

In the best case scenario your mates find you in the morning, if not, then at least you can start the day off fresh, with a plan of being found or of finding your way out.

S.E.R.E
(Survive, Evasion, Resistance, and Escape)

While S.E.R.E. may or may not relate to the Professional Citizen and wilderness survival, there is a preparedness mindset that crosses the boundaries between the military and everyday civilians that cannot be denied. Much of that has to do with:

THE WILL TO SURVIVE

That pure will to survive is a powerful force and is often the difference between making it and not making it. Straight out of the military field manual on S.E.R.E. we find this list on just the mental aspect of it.

Psychology of Survival
A. Be prepared:
*Know your capabilities and limitations.
*Keep a positive attitude. Lift yourself up.
*Develop a realistic plan.
*Anticipate fears.
*Combating psychological stress:
     -Recognize and anticipate existing "stresses."
     (injury, death, fatigue, illness, hunger)
     -Attribute normal reactions to existing "stresses."
     (fear, anxiety, guilt, boredom, depression, anger)
     -Identify signals of distress created by "stresses."
     (indecision, withdrawal, forgetfulness, carelessness, and propensity to make mistakes)
B. Strengthen your will to survive with:
*The Code of Conduct.
*Pledge of allegiance.
*Faith in America.
*Patriotic songs.
*Thoughts of return to family and friends.
C. Group Dynamics of survival:
*Take care of your buddy.

*Work as a team.
*Reassure and encourage each other.
*High morale is a result of group cohesiveness and well-planned organization:
  -Prevents panic.
  -Creates strength and trust in one another.
  -Favors persistency in overcoming failure
  -Facilitates formulation of group goals to overcome obstacles.
Five factors that influence group survival:
  -Enforce the chain of command.
  -Organize according to individual capabilities.
  -Accept suggestions and criticism.
  -Success often requires on-the-spot decision making.
  -Confidence is gained through knowledge and survival skill proficiency.

The author on a field exercise practicing camouflage and evasion

Another aspect of S.E.R.E. that crosses over is the need for putting together a survival kit.

PERSONAL SURVIVAL KITS

During the haste and confusion of survival, folks may become separated from their gear. A personal survival kit should be carried for such a contingency. The limiting factors for such kits are bulk and weight. There are three approaches to carrying personal survival kits: One is to scatter the items throughout clothing, so the weight and bulk is distributed across the body. The second way is to pack items in one pouch or container to make it easy to check for both location and contents.

The third method would be a combination of the first two. By making several small kits (fishing, medical, and repair kits) and scattering these sub-kits throughout clothing to accomplish one task. This also decreases the chance of losing the smaller items. The packing of these items, regardless of which approach is used, is also important. The containers must be small, sturdy, without sharp corners, and must provide protection for the kit contents. Some good containers are; plastic soap dishes, dental floss containers, film canisters, Band-Aid boxes, small leather or canvas patches, aspirin tins, heavy plastic bags, etc. If the container has a survival function such as a water container or signal mirror, so much the better. Small sharp items such as hooks, files, needles, razors and saw blades, are easily taped to a piece of thin cardboard, and carried in a wallet.

The following is list of items to choose from when designing a personal survival kit for survival emergency needs:

Medical/Health Needs:
-Sterile dressings
-Small bar of odorless soap
-Insect repellent
-Salt and sugar packets
-Water purification tablets/straws
-Chapstick/sunscreen

-Assorted Band-Aids
-Personal medications
-Tweezers
-Betadine pads
-Aspirin
-Moleskin
-Handiwipes
-Safety pins
Carrying/Holding Water:
-Plastic bags
-Plastic sheeting
-Prophylactics
-Heavy aluminum foil
-Surgical tubing
-Water tubing
-Ziplock bags
Body Protection:
-Large plastic bag
-Space blanket
-Evasion chart
-Floppy hat
-Wool watch cap
-Spare wool socks
-Bandana/neckerchief
-Neck scarf/cravat
-Gloves
-Sunglasses

PERSONAL SURVIVAL KITS
Camouflage Aid:
-Camouflage sticks
-Camouflage compact
-Burlap
-Camouflage compact Socks (disguise boot prints)
-Netting
-Cammie scarf
Heat/Fires:
-Matches (waterproof)
-Cotton balls

-Metal match
-Magnesium block
-Small candle
-Lighter
-Magnifying glass
-Petroleum gauze pads
Repairing/Improvising:
-Multi-bladed pocketknife
-Single edge razor blades
-Small sharpening stone
-Flexible wire saw
-Duct/electrical tape
-Para 550 cord
-Assorted needles
-Safety pins
-Safety wire
-Heavy thread
Navigation/Travel:
-Evasion chart
-Signal mirror
-Area topographic map
-Spare radio battery
-Compass
-Pen light
-Small pencil
-Whistle
-Small high quality monocular
Food Procurement:
-Fish hooks (assorted)
-Sinkers
-Bouillon cubes
-Nets
-Lures/spinner s/flies
-Snare wire

As an example of the on-body survival kit philosophy I took
the British smock loadout concept and minimized it to
work with a BDU field shirt.

Right bicep pocket:
-ferro rod w/striker
-lighter wrapped in duct tape
-wet fire tinder

Right chest pocket:
-knife
-Boo-boo kit
-red pen light

Left chest pocket:
-Compass w/mirror
-Map of AO

Left bicep pocket:
-write in Rain notebook w/pencil
-black sharpie
-chemlight with cord for a buzzsaw

The idea here with this is to have the bare survival essentials on your body in case you are separated from your kit. Most items are dummy corded in the pockets to prevent accidental loss. Essentially, think about what you would want on your body in an escape & evade scenario and set up your on-body kit accordingly.

Now, another example utilizing the survival kit philosophy is packing all the survival items in one, separate kit. This can be easily worn by itself or strapped to your pack. Here we put together such a kit to fit inside of a fanny pack.

Here is a breakdown of the contents...

-Tourniquet
-Pressure Bandage
-Shears
-Rolled Gauze
-Triple Antibiotic Ointment
-Burn Cream
-Hydrocortisone Cream
-Antihistamine
-Motrin
-Misc Bandaids
-Moleskin
-Swiss Army Knife
-Chemlight Buzzsaws for signaling
-Gutted 550 cord
-Whistle
-Button Compass
-Red LED light
-Collapsable Water Bottle
-Ziplocks (freezer bags) for water collection x2
-Iodine Tablets
-Lifeboat Matches

-Ferro Rod w/Striker
-Lighter wrapped in duct tape/jute twine
-Fatwood
-Tea Candle

So, as you can see it is its own, stand-alone kit to easily add to whatever you're rolling in. Is there redundancy in it when compared to items in an LBE or Chestrig, sure there is, but it's enough on its own to get you out of a tight spot and give you a fighting chance to survive and fight another day.

After all, at the end of the day you have one priority above anything else...

SURVIVE

Now, we have covered many aspects of S.E.R.E. in previous chapters such as shelters, water and food procurement, navigation, first aid, and more. But let's discuss the evasion aspect for a bit. I think this is highly overlooked within the preparedness community. We discuss in detail how to get found (as we did earlier in this book) but what if you don't want to be found. If you look at recent conflicts around the world, you will see cases where remaining hidden and possibly on the run are well documented. So let's look at the four main aspects of evasion.

Planning
Guidelines for successful evasion include:
*A positive attitude. You will see this info again.
*Use established procedures. For every outing part of your pre-trip planning is to set procedures for what to do if you get bounced from your camp or the outing goes horribly wrong.

*Follow your evasion plan of action. See above.
*Be patient and flexible. Flexibility is one of the most important keys to successful evasion. The evader is primarily interested in avoiding detection. Remember that people catch people. If the evader avoids detection, success is almost assured.
*Drink water. Do not eat food without water.
*Conserve strength for critical periods.
*Rest and sleep as much as possible.
*Stay out of sight.

The following odors stand out and may give an evader away:
*Scented soaps and shampoos.
*Shaving cream, after-shave lotion, or other cosmetics.
*Insect repellent — camouflage stick is least scented.
*Gum and candy — have strong or sweet smell.
*Tobacco — the odor is unmistakable.

Where to go? What is your plan? Meeting places in case things go sideways? Initiate evasion plan of action:
*Near a suitable area for recovery.
*Selected area for evasion (SAFE).
*To a neutral or friendly country or area.
*Designated area for recovery (DAR).

Camouflage
Basic principles:
*Disturb the area as little as possible. Make it hard for people to track you.
*Avoid activity that reveals movement to the enemy.
*Apply personal camouflage.

Use camouflage patterns
*Blotch pattern:
          -Temperate deciduous (leaf shedding) areas.

-Desert areas (barren).

-Barren snow.

*Slash pattern:

　-Coniferous (evergreen) areas — broad slashes.

　-Jungle areas — broad slashes.

　-Grass — narrow slashes.

*May use a combination of both.

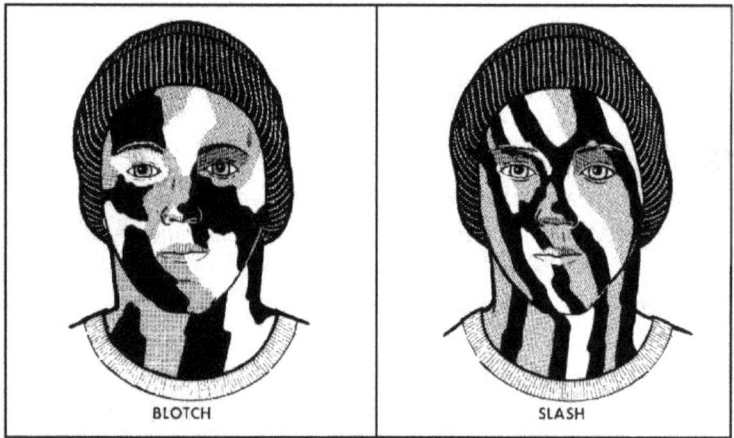

Examples of face camouflage.

Camouflage application:

*Face. Use dark colors on high spots and light colors on the remaining exposed areas (mask, netting, or hat help).

*Ears. The insides and the backs should have two colors to break up outlines.

*Head, neck (do not forget), and under chin. Use scarf, collar, vegetation, netting, or coloration methods.

*Give special attention to conceal light colored hair with a scarf or mosquito head net.

*Do not overlook hands, ears, neck, and body.

*Avoid unnecessary movement.

*Take advantage of natural concealment:

-Remember, foliage fades and wilts, change regularly.
-Change camouflage depending on the surroundings.
-Do not select all items from same source.
-Use stains from grasses, berries, dirt, and charcoal.
*Do not over camouflage.
*Remember when using shadows, they shift with the sun.
*Never expose shiny objects (i.e., watch, glasses, pens).
*Ensure watch alarms and hourly chimes are turned off.
*Remove unit patches, name tags, rank etc.
*Break up the outline of the body; "V" of crotch / armpits.
*When observing an area, do it from a prone and concealed position.

Shelters
*Use camouflage and concealment.
*Locate carefully:

Easy to remember acronym: BLISS.

B – Blend
L - **L**ow silhouette
I - **I**rregular shape
S – **S**mall
S – **S**ecluded location

Choose area:
*Least likely to be searched (drainage's, rough terrain, military crest, etc.) and blends with the environment.
*With escape routes.
*With observable approaches - do not comer yourself.

*Be wary of flash floods in ravines and canyons.
*Concealment with minimal to no preparation.
*Select natural concealment area.
*Consider direction finding (DF) when transmitting from shelter.
*Locate entrances / exits in brush and along ridges, ditches, and rocks to keep from forming paths to site.
*Ensure overhead concealment.

Movement
A moving object is easy to spot. If travel is necessary:
*Mask with natural cover (see Figure 1-2).
*Use the military crest.
* Restrict to periods of low light, bad weather, wind, or reduced enemy activity.

Do not be this guy.

*Avoid silhouetting
*Practice sporadic SLLS:
       -STOP at a point of concealment.
       -LOOK for signs of human or animal activity;
       (smoke, tracks, roads, people, vehicles, aircraft,
       wire, buildings, etc). Watch for trip wires or booby

traps and avoid leaving evidence of travel.
Peripheral vision is more effective for recognizing movement at night and twilight.
-LISTEN for vehicles, people, aircraft, weapons, animals, etc.
-SMELL for vehicles, people, animals, etc.
*Employ noise discipline, consider clothing and equipment.
*An important factor is breaking up the human shape or lines that are recognizable at a distance.
*Route selection requires detailed planning and special techniques (irregular route / zigzag) to camouflage evidence of travel.
*Some techniques for concealing evidence of travel are:
    -Avoid disturbing the vegetation above knee level.
    -Do not break branches, leaves, or grass.
    -Use a walking stick to part vegetation and push it back to its original position.
    -Do not grab small trees or brush. This may scuff the bark and create movement that is easily spotted. In snow country, this creates a path of snowless vegetation revealing your route.
*Pick firm footing, carefully place the foot lightly, but squarely on the surface avoiding:
    -Overturning ground cover, rocks, and sticks.
    -Scuffing bark on logs and sticks.
    -Making noise by breaking sticks (cloth wrapped around feet helps muffle this).
    -Slipping.
    -Mangling of low grass and bushes that would normally spring back.
*When tracks are unavoidable in soft footing, mask by:
    -Placing track in the shadows of vegetation, downed logs, and snowdrifts.
    -Moving before and during precipitation allows tracks to fill in.

-Traveling during windy periods.

-Taking advantage of solid surfaces (logs, rocks, etc.) leaving less evidence of travel.

-Brushing or patting out tracks lightly to speed their breakdown or make them look old.

*Do not litter. Trash or lost equipment identifies who lost it. Secure everything, hide, or bury discarded items.

*If pursued by dogs, concentrate on defeating dog handler.

*Obstacle penetration:

-Enter deep ditches feet first to avoid injury.

-Go around chain-link and wire fences. Go under if unavoidable.

-Penetrate rail fences, passing under or between lower rails. If impractical, go over the top, presenting as low a silhouette as possible

-Cross roads after observation from concealment to determine enemy activity. Cross at points offering

the best cover such as bushes, shadows, bend in road, etc. Cross in a manner leaving your footprints parallel (cross stepping sideways) to the road.

-Observe railroad tracks just like roads. Then align

body parallel to tracks and face down, cross tracks, using a semi-pushup motion; repeat for second track.

Notes:

CHAPTER 4
Mindset and Philosophy

Your mindset and emotions can be your best friend or your worst enemy when out in the field. Especially if you are just on the edge of your own ability. The best guys (or gals) aren't necessarily the most in shape or the most skilled. But rather they share a passion for this lifestyle with the ability to exert their will and to pay attention to both internal and external conditions.

Great Professional Citizens are the ones who remake themselves. They have the ability to strip away impediments from everyday life and can create a new character that is ready for the challenges ahead. Most are born with a certain internal fire that they have tempered with the realization that only an unsentimental view of themselves will show them where we need to improve and learn. Once we see the path to our goals, we must adhere to it regardless of setbacks and difficulties.

As we stated at the beginning of this book. It does make sense to emulate the greats in the outdoors, whether it be climbers, survivalists, soldiers, or the Mountain or Frontiersmen of our ancestors but do not look at their accomplishments. Instead, learn from their preparations. Their focus on the mental over the physical. At some point during any difficult situation that pushes you to your limit, the only strength that matters is the mental one.

But in order to do all of this, you must know who you are, want it is that you want, and what pushes you. It all begins with self-understanding for building self-control. Practicing self-discipline creates habits of the mind that start to kick in automatically as your own experiences begin to grow. With experience will come self-confidence, which is required for survival anywhere in the world.

Self-Assessment: As you decide to start on this journey, you need to assess your own personality. Are you more of an engineering minded person or are you more artistic minded? Some folks will approach outdoor skills and survival like an engineering problem. There is a checklist of things that need to be done in order to solve that problem. These folks will train in a quantifiable way, fine tuning caloric intake, meals, gear weight, organization, weather, and more. Every aspect of their outing is planned.

Other folks will act like the artist. They look at a problem and, for some reason, just know how it is going to go. These folks have this firm belief that it is just going to work out. Planning tends to be casually done. In simple terms, they can just load a pack with gear, food, and fuel and upon shouldering just seem to know if it will be enough or not.

The best case is to be a little bit of both of these types of people. Find out which one speaks to you the loudest and go with it.

At the beginning of this book, we mentioned the "crawl, walk, run" philosophy for learning and training. This journey is a constant evaluation of your experiences, strengths, weaknesses, and abilities. If you are new to this and just spent a night out in the woods camping don't think that you can go thru hike the Appalachian trail the following week. Your experience is insufficient at this point. However, your goals are important, and you should strive to reach them. Lofty goals are what push us to do great things. So, recognize your goals and push through to reach them, even running towards them every now and then. But they still belong in the future at this point.

A trap I have fallen into in the past is accidentally succeeding on a climbing trip that was way above my ability level and then thinking that I could tackle a similar climb, when in reality I got lucky. It wasn't skill that did it, it was luck in having good weather and a good partner. You will need to learn to recognize when you get lucky meeting a challenge. If you don't learn to understand this now, it will lead to a casual attitude that will lead to carelessness and failure that can impact you and others. Have respect for the process to attain the skills to do the hard things.

Sanguine		Choleric	
Strengths	**Weaknesses**	**Strengths**	**Weaknesses**
-sociable	-impulsive	-ambitious	-agressive
-charismatic	-chronically late	-passionate	-domineering
-outgoing	-shameless	-leader-like	-inflexible
-confident	-forgetful	-focused	-impatient
-warm-hearted	-a compulsive talker	-efficient	-rude and tactless
-pleasant	-too loud	-practical	-argumentative
-lively	-sometimes too happy	-good at planning	-unable to relax
-optimistic	-distractible	-good at problem solving	-uncomfortable around emotion
-a fun lover	-not interested in following through with tasks that are boring	-confident	-low on empathy
-spontaneous		-motivating	-discouraged by failures
-a preventer of dull moments	-self-absorbed	-a delegator	-too busy for people
-a quick apologizer	-an exaggerator	-usually right	-impatient
-an easy friend maker	-someone who appears unauthentic	-great in an emergency	-a leader who demands loyalty

Phlegmatic		Melancholic	
Strengths	**Weaknesses**	**Strengths**	**Weaknesses**
-relaxed	-sometimes shy	-thoughtful	-obsessive
-quiet and calm	-fearful of change	-considerate	-too cautious
-content with themselves	-prone to laziness	-cautious	-prone to depression
-kind	-stubborn	-organized	-prone to moodiness
-consistent	-passive-aggressive	-an excessive planner	-perfectionistic
-a steady and faithful friend	-indecisive	-schedule oriented	-pessimistic
-accepting	-permissive	-detailed	-difficult to please
-affectionate	-not goal oriented	-highly creative in poetry, art and invention	-deeply affected by tragedy
-diplomatic	-unenthusiastic	-independent	-a person with tunnel vision
-peacemaking	-too compromising	-good at preventing problems	-sometimes a procrastinator
-rational	-undisciplined		-discontent with themselves and others
-curious	-sarcastic		-prone to play the martyr
-observant	-discouraging		
-an easy friend maker	-non-participative		

hidingplaceblog.blogspot.com

Another important factor is your overall temperament. There are 4 types of temperaments and looking into them can help you in understanding your type and this can help guide you into how you are most likely to react in different situations. Combine your abilities with your temperament and you can have a recipe for success.

Don't be afraid to challenge yourself and look for ways to maximize your strong points, and where you may be mentally comfortable and satisfied. Go experiment with different methods, conditions, and gear and see how you perform physically and mentally. Push yourself to see what your limits might be. I routinely do this when it comes to weather conditions. Picking freezing temps or less than ideal conditions, I will make a field exercise out of it to see how I handle the extremes. Maybe it is using a minimal sleep kit on a cold night, maybe it is a rainy night out in a poncho shelter, or it could even be a 24-hour exercise with using only what you can carry in your pockets. Whatever it is you can only truly know yourself by pushing yourself. Do not use other folks' accomplishments to judge your own by.

Overcoming Fear: No one is exempt from experiencing fear. Those who say they aren't afraid aren't being honest with you, or themselves for that matter. Although seasoned climbers, outdoorsmen, and soldiers feel comfortable in scary situations, they know to fear only that which they have no control over. You can fear something but not be paralyzed by it. The veterans know how to direct that fear into some productive action. Your mind is what produces the fear, so the need to direct that into something controllable is a mental exercise in using it as a source of energy.

There is a famous climber, Mark Twight, who had a great thought on dealing with fear. He said:

*"Nobody controls a situation in the mountains. It is vanity to imagine one can. Instead, grow comfortable with giving up control and acting within chaos and uncertainty. Attempting to dominate constantly changing circumstances in the mountains or to fight the loss of control serves only to increase fear and multiply its effects. **Embrace the inherent lack of control and focus on applying skills and ideals to the situation.**"*

Sometimes with fear you need to break it down into small bits to tackle and overcome. Small goals to achieve a bigger goal. An example might be getting caught out on an unexpected overnighter in the cold and rain. Instead of retreating out of fear of the cold and wet, focus on the needs of getting through it. Remember the survival priorities from earlier? Shelter, fire, water. You can't control the weather, but you can control tackling those survival needs. So focusing on what you can do will get you through those fearful situations.

Failure to have self-discipline and letting your fear run amuck will lead to an out-of-control panic. I have panicked in the mountains. It led to a retreat in New Hampshire's Presidential Range on a winter traverse. At the time I was a bit inexperienced in dealing with fear, but instead of relying on my skills and familiarity with the peaks, I panicked because of a pending storm. I had the right gear, the right skill set, and even the right experience to deal with it but, instead, I panicked and retreated to seek comfort in a lodge. I let my fear hold me hostage. Not only was I disappointed in myself, but my climbing partners were disappointed too. A couple years later I had learned to control my fear and successfully did the winter traverse of that mountain range.

Failure:

My kids have asked me how many mountains I have climbed, my response to them has always been:

"It isn't the number that I have been successful on but the number that I have failed on. That number is greater than the ones I have stood on top of, but I learned more from those I failed on.

Those experiences hold great value to me in my overall level of training and ability. It took me several years to go from being a goal-oriented climber to being an experience-based climber. My goal became the experience, not the summit, and I became a much better climber because of it. This easily translates to The Professional Citizen. Do not be the one to value summits over experience, too many folks who do have been killed doing so. No one is immune to this. But folks who want to learn, who want to be around long enough to experience the journey and do good things, that other folks and family can trust, must learn to fail and when to fail.

Learning to fail is tough, we don't let people fail anymore it seems. Everyone gets a trophy just for being there. We have failed an entire generation because we don't teach failure anymore. It is a tough lesson to learn. And because of that folks don't know how to deal with it. Even the greats fail, but they learn to fail in a positive way, they fail by trying. Failure is part of the process, probably the most important part of it.

One aspect of failure is learning when to turn back and retreat. Do this before you lose your ability to decide what is going to happen. That only comes with experience in the field doing the things to get better. Give it a good fight, don't panic, and keep a good attitude about it. The will to live is programmed into all of us, although some folks' will is stronger than others. How strong is yours? Will you just give up without a fight? There is only one way to know.

Positive Mental Attitude (PMA):

I am a glass half full kind of guy. I always try to look at the positive in a situation. It keeps my head clear and allows me to look for solutions instead of dwelling on the negatives of the situation. A bad attitude will guarantee failure, plain and simple. You need to want to be where you are. If you aren't in it 100% then don't do it, period.

About 30 years ago I took a mountaineering course from a well-known climber by the name of Alex Van Steen. He was the first person that I ever heard the phrase "Positive Mental Attitude" from. It was day one of the course and boy did that one phrase resonate with me. He encouraged forward thinking in problem solving, creativity in action, and most of all pure enjoyment in what you are doing. It was from this that, as I got older in life, I started using my own phrase on having a Positive Mental Attitude being:

"I may not be the fastest, nor the strongest anymore, but I will always be the guy with a shit eating grin on my face because I enjoyed it more than most."

To me, this embodies having a Positive Mental Attitude.

Negativity breeds negativity. Have you ever been in a room full of folks that complain, and you find yourself complaining too. Does anything ever get accomplished by this? No, of course not. Same goes for out in the field. Yes, things may suck, yes it may be hard, hell the situation may be dire, but a bad attitude will not fix it. Strive to be the positive light. It will breed a good attitude from those around you and solutions will come from it, then things may suck just a little bit less.

Embrace The Suck:

In the world of climbing, we had used the phrase "The Art of Suffering" as a way to say, "Embrace the Suck". It's the same philosophy, just different people.

But, in application, it is a mindset of pushing through conditions, both mental and physical, beyond what you may normally find acceptable. It is a way to lean into the suffering and to get comfortable being very uncomfortable. This doesn't happen overnight and takes not just experience, but it also requires a state of mind in literally embracing the challenge ahead. This can be applied to not just the field, but everyday life as well.

Ever hear the phrase "you don't have to like it, but you have to do it"? Well, that goes along the same lines as "Embrace the Suck" too. In the field we don't always get the perfect day, in fact it never really goes as planned. Rain, wind, snow, heat, difficult terrain, mud, whatever it may be, you need to be able to get comfortable being out in it. Every winter I do a solo cold camp where I wait until an especially sucky, cold, and snowy day and I hit the woods with minimal kit for a 24-hour outing. My wife will shake her head telling me I'm crazy, my son will tell me "no" when I ask him to go, but I do it as a test. There have been times that I don't sleep at all, there have been times that the stars line up and I slept like a baby, and other times I have willingly stayed up listening to the coyotes call to each other all night while I try to find out where the packs are. This yearly sabbatical to test my grit is my favorite field exercise of the year. It embodies the willingness to suffer. It is a mental test for me. Experience has taught me that physically I can do it, no issues there, but it is the mental test that intrigues me every year.

So herein lies the challenge of this philosophy, do you have what it takes to push through adversity? Physically, emotionally, and mentally? How do you really know? This is why we preach getting out into the field. Create scenarios and exercises to work on. Do it with full kit and then do it with minimal kit. Do it in bad weather and in good weather and everywhere in between. Push yourself until you don't think you can take another step or another moment and then take that next step. You will be surprised at just how far you can go if you can mentally accept the discomfort you're going to feel.

Last year I challenged myself to do a Norwegian Ruck March. I have spent a lifetime carrying a pack and trudging through difficult terrain, but it had been a number of years since I really felt as though I had pushed myself. So, I was prepared to "Embrace the Suck" and throw myself back into the ring, so to speak.

So here is a little background on the Norwegian Ruck March:

The Norwegian Foot March, also known as Marsjmerket, is an armed forces skill badge earned after completing an 18.6-mile foot march while carrying a 25-pound rucksack in under 4.5 hours. The march was first held during World War I in 1915 to test the strength and endurance of the Norwegian military while exposing them to conditions they could expect in combat. The goal was to move large groups of soldiers across long distances with fighting loads and still be combat effective.

Up until a few years ago it was only open to service members but now us civilians can take part in this test of endurance. So, I signed up and did it. 97 people participated, I was 1 of 3 civilians and the second oldest overall. My age group (which I was the only one) had to do it in 4 hours and 40 minutes. I came in at 4 hours and 27 minutes (which was the time frame for the 20- to 34-year-olds). A representative from the Norwegian Ministry of Defense was present to hand out certificates and medals for those who hit the standard.

I'm not going to lie, it was hard. We started off in the cold rain, then it got windy, but I just put my head down and put one foot in front of another. At mile 12 was when I

started to feel it. My knee started to tighten up, but I knew I could do it. One by one I passed folks half my age suffering, slowing down, even sitting on the side of the trail. But I just trudged along with that "Positive Mental Attitude" we spoke of earlier. I was just happy to be there, happy to be in the moment knowing full well that I was going to succeed at it.

I hadn't done anything that difficult in quite a while, but even though I'm not the fastest nor the strongest anymore, I still probably enjoy the journey more than most! Because that was really what doing this was all about, the journey!

So, as you can see, experience leads to a heightened mental toughness that one only gets by putting themselves out there. Doing the hard work now knowing that when the time comes, you will be better prepared to deal with the suck.

Notes:

Chapter 5
Putting it all Together

So, as we wrap this up, what does this all even mean to us? Why do I feel it is so important to be well versed and practiced in these basic skill sets and have a thorough understanding of outdoor fundamentals? It is really a culmination of a lifetime spent outdoors working to perfect my own craft and ways to apply it all. As I have gotten older, I have understood it more and more as not just being something that we do, but rather that it is just something that we are.

A few years ago, I finally realized that no one is coming to rescue us. It is on each one of us to become the best versions of who we can become to get ready for it. As I alluded to early on, we spend a lot of time emulating what we see on social media and in print media, but very few experience the process. The long and hard work that must be put into preparing oneself to be the one that will be his family and friends' protector. We should all strive to be the one. It's our duty to do so. But these skills and philosophies will not only have us better prepared in the long run, but they are useful in our everyday adventures.

How many times do we see the aftermath from natural disasters, wars, and accidents? Do you ever see the video footage or read the stories and ask yourself "Why weren't they prepared? Why didn't they do this or why didn't they do that?". Well herein lies one of the basic reasons to get prepared. To start training to be the one that will help keep your family safe and to be the one that they rely on.

You will be your own rescue, your own QRF team, your own supply train, and so much more. Now is the time to network with like-minded people in your area. Strengthen the bonds you have with those you already know. Quit buying gear and start learning to use it. We continue to chase problems with gear instead of training and gaining experience to tackle those problems. This is where we, as a community, need to do a better job helping one another. If you have a group of friends who share these interests and goals, then start putting together training weekends with them. Each of you has things that they are good at in the field, start cross training with each other. If one of you is great at fire making, then pick a day or weekend and teach everyone else what you know. Same goes for first aid, navigation, shelter building, cooking, climbing, shooting, hunting, and so much more. Come up with scenarios and play them out. Make it fun but make it educational. You can then sit around the fire and breakdown what went right, what went wrong, and what needs to be worked on next. Use this to inspire and encourage one another to get better.

This path is not easy, you will make mistakes and you will fail at times, but it will be so worth it in the long run. Do not be afraid to make mistakes and to fail. It will breed success down the road, and you will be better because of it.

My hope is that you find encouragement and direction with this and that it leads you to being much more prepared, not just for yourself, but for those around you. It is worth it!

Notes:

Checklists, Diagrams, and Forms

In the section below you will find helpful planning forms, equipment checklists, and diagrams that can help you with putting kits together, planning for training exercises and outings, and just useful information for getting yourself on the way to overall preparedness.

The Ten Essentials:

1. Map/Compass
2.Headlamp/Flashlight
3. First Aid
4. Water
5. Knife
6. Food
7. Extra Clothing
8. Rain Gear/Shelter
9. Fire Starter/Ignition Source
10. Cordage

Emergency Forms:

HIKING PLAN

Complete this form before departing on a hike and leave it with a reliable person who can be depended upon to notify authorities in case you do not return as scheduled. A word of caution: In case you are delayed and it is not an emergency, inform those with your hiking plan of your delay in order to avoid an unnecessary search!

1 Names of person filing this plan:_____
Telephone #: (_____) _____ - _____

2 Name(s) of others on hike: Age: Address:

_____ ____ _____
                    Telephone #: (_____) _____ - _____

_____ ____ _____
                    Telephone #: (_____) _____ - _____

_____ ____ _____
                    Telephone #: (_____) _____ - _____

_____ ____ _____
                    Telephone #: (_____) _____ - _____

_____ ____ _____
                    Telephone #: (_____) _____ - _____

_____ ____ _____
                    Telephone #: (_____) _____ - _____

3 Radio ☐ Yes ☐ No   Type:_____   Call sign: _____   Frequncies _____

4 Trip plans
Leaving from: _____   Going to: _____

Route details:_____

Departing on:_____ / _____ ☐ am ☐ pm   Returning:_____ / _____ ☐ am ☐ pm
          date   time   date   time
And, in no event, returning later than: _____ / _____ ☐ am ☐ pm
                              date   time

5 Alternate route if bad weather is encountered :_____

_____

6 Description of automobile: _____

7 Make:_____ Color:_____ License #:_____ Where parked: _____

If not returned by: _____ / _____ ☐ am ☐ pm
Call: date   time
Local authority:_____   Telephone #: (_____) _____ - _____

Tactical Combat Casualty Care (TCCC) Card

TACTICAL COMBAT CASUALTY CARE (TCCC) CARD

BATTLE ROSTER #: _____

EVAC: ☐ Urgent ☐ Priority ☐ Routine

NAME (Last, First): _____ LAST 4: _____

GENDER: ☐ M ☐ F DATE (DD MMM YY): _____ TIME: _____

SERVICE: _____ UNIT: _____ ALLERGIES: _____

Mechanism of Injury: (X all that apply)

☐ Artillery ☐ Blunt ☐ Burn ☐ Fall ☐ Grenade ☐ GSW ☐ IED
☐ Landmine ☐ MVC ☐ RPG ☐ Other: _____

Injury: (Mark injuries with an X)

TQ: R Arm		TQ: L Arm
TYPE:		TYPE:
TIME:		TIME:

TQ: R Leg	TQ: L Leg
TYPE:	TYPE:
TIME:	TIME:

Signs & Symptoms: (Fill in the blank)

Time				
Pulse (Rate & Location)				
Blood Pressure	/	/	/	/
Respiratory Rate				
Pulse Ox % O2 Sat				
AVPU				
Pain Scale (0-10)				

DD Form 1380, MAR 2014 — TCCC CARD

BATTLE ROSTER #: _____

EVAC: ☐ Urgent ☐ Priority ☐ Routine

Treatments: (X all that apply, and fill in the blank) Type

C: TQ- ☐ Extremity ☐ Junctional ☐ Truncal

Dressing- ☐ Hemostatic ☐ Pressure ☐ Other

A: ☐ Intact ☐ NPA ☐ CRIC ☐ ET-Tube ☐ SGA

B: ☐ O2 ☐ Needle-D ☐ Chest-Tube ☐ Chest-Seal

C:	Name	Volume	Route	Time
Fluid				
Blood Product				

MEDS:	Name	Dose	Route	Time
Analgesic (e.g. Ketamine, Fentanyl, Morphine)				
Antibiotic (e.g. Moxifloxacin, Ertapenem)				
Other (e.g. TXA)				

OTHER: ☐ Combat-Pill-Pack ☐ Eye-Shield (☐ R ☐ L) ☐ Splint
☐ Hypothermia-Prevention Type: _____

NOTES:

FIRST RESPONDER
NAME (Last, First): _____ LAST 4: _____

DD Form 1380, MAR 2014 (Back) — TCCC CARD

195

IFAK (Individual First Aid Kit)

With the threats currently facing US Citizens, the need for basic tactical emergency medical skills and equipment cannot be overstated. Having the basic skills and necessary equipment to treat a life-threatening penetrating injury such as a gunshot wound is crucial. It only takes between two and four minutes to bleed to death from an injury to a major artery. It will be your personal skills and available equipment that will likely be the difference between life and death. You must be mentally prepared to come out of a critical incident with a life worth living mindset, not a mere survival mindset where your quality of life has been irreparably damaged. Training and preparing to prevail and not merely survive is a responsibility that falls squarely on each of us.

Assembling your own IFAK is relatively simple and inexpensive and should include (as a minimum) all of the items necessary to treat life threatening injuries caused by penetrating trauma such as gunshot or stab wounds.

Basic Items
-IFAK Pouch
-Combat Tourniquet
-Israeli Bandage
-Quick Clot/Combat Gauze
-Elastic bandage
-Gauze Bandages
-Self stick medical wrap
-Triangular Bandage
-Surgical gloves

Advanced Items
-Two 14-gauge Angiocath Needles (3" or greater in length)
-One (size 26-28) French Nasopharyngeal Airway w/small pack of surgical gel (NPA)
-Two Chest Seals such as Asherman or Hyfin

The Professional Citizen Personal Equipment List

*Rifle
>   -AR or AK would be the norm but any 5.56 or 7.62
>   semi auto would be of benefit to any group!
>   -7 magazines for a total of 210 rounds

*Sidearm
>   -3 magazines (personal preference for what you
>   carry)

*Extra Ammo

*Weapons mounted light for rifle

*Firearm cleaning kit

*Fixed blade fighting knife

*Folding knife/Multitool

*Tomahawk/Hatchet
>   -I wouldn't dismiss the utility of this item, whether
>   as a weapon or tool for creating a shelter or even a
>   creative trap/diversion for the enemy...

*Shovel/Entrenching Tool
>   -Once again, the utility for creating a fighting
>   position, digging a dakota fire trench, and digging a
>   cat hole...

*Armor
>   -If you plan on getting into a gunfight then I think
>   armor is a must have, I will include a helmet...
>   Head protection can be anything from ballistic
>   protection to a bump helmet, protecting your
>   thinking tool may be one of the most important
>   things you do!

*Battle Belt
>   -1st line layer with sidearm, mag pouches for both
>   sidearm and rifle, knife, IFAK,
>   flashlight/headlamp, compass, map, water, fire
>   stick, etc....

*Chest Rig/Plate Carrier/LBE/Load Bearing Vest...
>   -2nd line layer with armor, additional mag pouches,
>   comms, and water...

*Day Pack/Sling Bag/Hydration Pack

-There are styles that will attach directly to LBE's and Chest Rigs or Plate Carriers)

*Operating Pack (30-40 Liters)

-3rd line layer (This would be for extended operations... Worn in place of the hydration carrier or day pack. Room enough for group gear, personal gear, spare ammo, bedroll, stove, rations, water, etc...)

-(This pack would vary in size for your use... Most packs in this category will have hydration sleeves built in, compression straps for snugging it down if it isn't filled to capacity, removeable internal frames, MOLLE straps, plus extra compartments for organization... Military surplus is great, plus some private companies specialize in operator style packs such as Eberlestock, Karfu, Hill People, Tactical Tailor, Mystery Ranch, Arcteryx, etc...plus there are many "tactical" styles on the market (Condor, Rothco,, etc..) that will definitely suit these needs!

*Communications Equipment

-Multi-band, 2-way radios would be the norm, Baofeng (All around #1), Motorola, etc...

*10 Essentials

-1st aid, knife/paracord, map/compass, fire starting equip, water, food, shelter, warm clothes, rain gear, flashlight/headlamp...*(A Minuteman should never be without these 10 items)*

*First Aid (IFAK)

-Tourniquet, clotting sponges, Israeli Bandages, triangular bandage, SAM splint, eye bandages, burn pads, nitrile gloves, tape, saline syringes, antibiotic ointment, blister care/moleskin, aspirin, booboo band aids, suture kit, antihistamine, trauma shears, sharpie, wet wipes, alcohol prep wipes, etc...

*Stove and Fuel
> -Small canister fuel option like the Jet Boil series or MSR pocket rocket are quick, quiet, and efficient! Multi fuel options are great for heavy use, 3rd world locations, and extreme cold climates. They will burn everything from white gas to both unleaded and diesel fuel, kerosene and even jet fuel! Options like the MSR XGK are workhorses, just heavy and bulky... Or small emergency stoves like the Esbit and solo stoves that use fuel tablets or natural combustibles are a great lightweight, simple option! Remember, we are not cooking gourmet meals, simply boiling water (maybe even melting snow) for a dehydrated meal or quick coffee and tea!

*Water/Water Treatment
> -This is obviously your life source, 3 liters per day per person is a good rule of thumb...Hydration bladders, collapsible water bottles, Nalgene or Klean Kanteens, take your pick! No excuse not to have water....
> -Filters like the Sawyer mini, Lifestraw, or pump style like the Katadyn Hiker or treatment options such as iodine or chlorine tablets

*Shelter
> -From a basic survival tarp, Bivy sack, space blanket to a hammock to a traditional tent, this choice will vary on season, area, and mission... Keep in mind, you're not out to set up a cozy camp and tell ghost stories... Quick setup, quick takedown, low profile and low visibility should certainly be taken into consideration!

*Sleeping Bag/Pad
> -Poncho liner, Snug sack, fleece/wool blanket, LW sleeping bag, etc... Closed cell foam pad, lg enough to go from you hips to your shoulders. Depends on the season, AO, and mission...

*Food

 -Energy bars, GORP (trail mix), nuts, jerky, SPAM, MRE's, dehydrated meals, etc... Easy to eat on the go items... Avg of 2,000 calories a day is the norm so set yourself up for being able to sustain yourself...

MISCELANEOUS GEAR

*Gear repair kit

*Personal hygiene kit

 -TP, Wet wipes, sanitizer, sm pack towel, foot powder, bug repellant, sunscreen, etc...

*Chem sticks

*Flares/Strobes

*Solar Charger

*GPS

*Burner Phone

*Optics

 -I would include Binoculars, Monoculars, NVG's, Therms, etc in this category.

*Protective Eyewear

*Camo Netting

 -For hideaways, sniper veils, encampments, etc...

*Climbing Rope, Webbing, Carabiners

 -(100' rope & 25' of webbing for a Swiss Seat and assault line)

*Notebook/Pencil/Sharpie

 -Taking notes while on patrol, sketching enemy camps, details of movements, etc.

CLOTHING

(This is a full overview of clothing and layers and what the "Professional Citizen" chooses would be based on the season, AO, and mission requisites)

*Underwear (quick dry)

*Long Underwear (Poly Pro or Merino Wool)

*Pants (This will obviously vary, depending on climate and season)

*Short Sleeve Shirt
*Long Sleeve Shirt
*Insulating layers (such as power stretch fleece tops and bottoms, LW synthetic filled pullovers or jacket, Down jackets/parkas for extreme cold climates/conditions)
*Outerwear (Waterproof/Breathable shell jacket and pants, Softshell technology is huge today and is almost 100% WP and a lot more breathable than Gore-Tex, plus the fabric is much quieter)
*Socks (Synthetic or wool with optional liners to help with blisters and foot rot)

> -I'm a huge supporter of the 3 pairs of sock rule, 1 on your feet, one pair that you just had on, and one pair to change into and you rotate!

*Gloves (Working gloves (tactical shooting gloves) that you can manipulate all your gear with are a huge bonus, plus lightweight liners, and warmer gloves/mittens (they make mitts with trigger fingers) for colder conditions)
*Headwear (Wide brimmed Boonie hat, billed cap, Stocking cap or watch cap, and a lightweight balaclava) (helmet is listed above with the gear)
*Footwear (Proper fitting boots are of the utmost importance, choose something with ankle protection but are relatively lightweight! Arch support is a must, remember, your feet are supporting not just your body weight but now an additional 30-50 plus lbs...

Patrol Kit Check List

LBE packed to work for a 1-to-2-day foot patrol.

-Gloves
-Smoke, large black x2,
-Smoke, small red x1, green x1
-AR mags x6
-Sheath knife
-Radio
-Rite in rain notebook
-Sharpies/pencil
-Laminate map
-Compass
-Water bottle 1 liter x2
-Nesting mug w/lid stainless steel x2
-Esbit stove with fuel cubes
-Spork
-Sawyer Mini filter w/16 oz bladder
-Iodine tablets

-Multitool
-Signal panel
-Rescue mirror
-Whistle
-Chem lights x2
-Spare batteries
-Bandana
-NVG's
-Binoculars 10x25
-TQ
-IFAK

In Butt pack:
-Sit pad
-Basha with bungees/pegs
-Rain jacket
-Ration pack
-Rifle cleaning kit
-Face camo
-Neck gator
-Spare socks
-Wet wipes
-GP pouch including
*headlamp
*cordage
*lighter/Ferro rod
*fire starter
-Contractor bag/space blanket

RATION KIT (example)

BREAKFAST:
-Quaker Oats Peanut Butter Bars (250 calories each X3)
-Coffee (zero calories X3 servings)

SNACKS/LUNCH:
-Nut Clusters w/Blueberries & Raspberries (420 calories each X3)
-Dried Fruit Mixes w/Cashews (340 calories each X4)
-Granola Bars (Variety 200 calories each X9)
-Pepperoni Beef Sticks (90 calories each X3)

DINNER:
-Ramon Noodles (380 calories per package X3)
-Beef Jerky (80 calories each X4)

Total Calories per day equals 2,160 (or 6480 calories for a 3-day package) ... Everything in this kit was found at the local Dollar store so it can easily be done on a budget.

This ration pack covers 1 person for 3 days' worth of food in the field. This pack allows for 1 hot meal a day but can be tailored to the end user if a hot meal is not an option.

Combat Service Support

CSS serves 4 different functions:
-Offensive situations will require as much support as far forward as possible. Ready resupplies as ammunition, weapons maintenance, and first aid will be critical at this level along with communication needs and up to date intelligence.
-Defensive situations require additional ammunition and medical but will also require additional support in the way of fortification material and manpower for help in turning a defensive posture into an offensive one.
-Stability operations can serve multiple purposes, from long term supply of forward operating teams to emergency and humanitarian efforts to peace enforcement missions which may require combat.
-Support operations most critical task is CSS... The Professional Citizen would offer and provide support to civilian agencies in time of humanitarian need and respond to civil unrest, emergencies and natural disasters. Being able to react quickly and efficiently to help with search & rescue efforts, medical needs, and additional security would be a valuable asset.

Needs of The Professional Citizen
-Sustaining group operations is a challenge for CSS planners and organizers. Each situation requires some combination of theater and contingency CSS. Todays CSS planners apply their knowledge of conventional CSS operations to meet their specific needs. The fundamentals of contingency CSS would apply to most group needs.
-The nature of such operations frequently imposes stringent operations security (OPSEC) requirements on the CSS system. Certain situations would be extremely sensitive and require compartmentalization of their support to avoid compromise. Supporting CSS planners ensure OPSEC within their own activities.
-Modern Minuteman units are comparatively small and will serve and participate in a varied number of roles and

missions. From long range reconnaissance patrols (LRRP) to quick reaction neighborhood response teams, to coordinated assistance with local civilian governments and LEO's they will consume few critical combat supplies (Class I, bulk Class III, and Class V). However, they use special operations-peculiar and low-density items of standard and nonstandard configuration. The solution to these CSS requirements is area-specific and mission-dependent.

Modular designed/mission specific packages will be the most efficient way to support one's group. Most groups would have an initial deployment package (IDP) that would be there basic kit with mission oriented gear and a follow-on package (FOP) would be staged and ready to supplement the mission at a predetermined point with request from the group!

Classes of Supply
I. Subsistence (food/rations), gratuitous health and comfort items.
II. Clothing, individual equipment, tentage, organizational tool sets and kits, hand tools, unclassified maps, administrative and housekeeping supplies, and equipment.
III. Petroleum, fuels, lubricants,
IV. Construction materials
V. Ammunition
VI. Personal demand items (such as health and hygiene products, soaps, and toothpaste, writing material, snack food, beverages, cigarettes, batteries, and cameras).
VII. Major end items such as mobile machine shops, and vehicles.
VIII. Medical materiel including repair parts peculiar to medical equipment.
IX. Repair parts and components
X. Material to support nonmilitary programs such as agriculture and economic development (not included in Classes I through IX).
Miscellaneous. -Water, salvage, and captured material.

SALUTE Report:

The standard format for reporting enemy information is the SALUTE report. SALUTE is an acronym that stands for Size, Activity, Location, Unit identification, Time, and Equipment. The report body should be brief, accurate, and clear.

SALUTE

- S: Size - report the number of personnel, vehicles, aircraft, or size of an object.
- A: Activity - direction of movement, troops digging in, artillery fire, type of attack, NBC activity
- L: Location – Where? Include grid coordinates or reference from a known point
- U: Unit - Report the enemy's unit. If the unit is unknown, report any distinctive features (uniforms, patches or colored tabs, headgear, vehicle identification markings)
- T: Time – When observed, not the time you report it.
- E: Equipment - Report all equipment associated with the activity (weapons, vehicles, tools). If unable to identify the equipment, provide as much detail as you can.

TSP 301-371-1000
0903, Phase II, OCS

SITUATION Report (SITREP)

The Situation Report (SITREP) is a form of status reporting that provides decision-makers and readers a quick understanding of the current situation. It provides a clear, concise understanding of the situation—focusing on meaning or context, in addition to the facts.

SITREP CH 3

SITREP FORM

(To pass along brief reports about a developing situation, and the dangers it may pose to you, your community, or to the region.)

1. FROM: *(Sender)*	2. TO: *(Recipient)*	3. PRECEDENCE:
4. Current DTG: *(YYYYMMDD-HHMMZ Use UTC)*		5. Incident Number: *(YYYYMMDD-HHMMZ Use UTC))*
6. Expiration: *(YYYYMMDD-HHMMZ Use UTC)*		7. Location: *(Lat/Lon, Grid Square, City)*
8. Incident Status:	9. Size and Scope:	10. Overall Hazard:
11. Current Weather:	12. 48 hr Weather:	13. Infrastructure:
14. Political:	15. Civil:	16. Communications:
17. Remarks:		

End of Report
AFP-110 REV 20150612

208

OBSTACLE Report

OBSTACLE REPORT (also referred to as a BLUE 9 report).

This is used to report threat or unknown emplaced obstacles to other friendly units or your leadership. Report all pertinent information using the following format:

Line ALPHA: Type of obstacle or obstruction.

Line BRAVO: Location, using grid coordinates. For large, complex obstacles, send the coordinates of the ends and all turn points.

Line CHARLIE: Dimensions and orientation.

Line DELTA: Composition.

Line ECHO: Threat weapons influencing obstacle.

Line FOXTROT: Observer's actions.

S.E.R.E
(Survive, Evasion, Resistance, and Escape)
QUICK REFERENCE CHECKLIST

DECIDE TO SURVIVE!

S – Size up the situation.
   Physical condition
   Adequate water intake
   Injuries, Illness
   Food
   Surroundings
   Equipment
U – Use all your senses, slow down and think.
R – Remember where you are.
V – Vanquish fear and panic.
I – Improvise and improve.
V – Value living.
A – Act like the natives.
L – Live by your training and experience

1. IMMEDIATE ACTIONS
   a. Assess immediate situation.
   b. Take action to protect yourself from NBC hazards.
   c. Seek concealment.
   d. Assess medical condition; treat as necessary.
   e. Sanitize uniform of potentially compromising information.
   f. Sanitize area, hide equipment you are leaving.
   g. Apply camouflage.
   h. Move away from initial site; zigzag pattern recommended.
   i. Use terrain to advantage; communication and concealment.
   j. Find hold-up site.

2. HOLD-UP SITE

    a. Reassess, treat injuries, inventory equipment.
    b. Review plan of action; establish priorities.
    c. Determine your current location.
    d. Improve camouflage.
    e. Focus thoughts on task(s) at hand.
    f. Execute plan of action. . .STAY FLEXIBLE!

3. CONCEALMENT

    a. Select a place of cover and concealment providing:
        (1) Adequate cover; ground and air.
        (2) Safe distance from enemy positions and lines of communication. (LOCs)
        (3) Listening and observation point.
        (4) Multiple avenues of escape.
        (5) Protection from the environment.
        (6) Possible communication/signaling opportunities.
    b. Stay alert, maintain security.
    c. Drink water.

4. MOVEMENT

    a. Travel slowly and deliberately.
    b. Do not leave evidence of travel, use noise and light discipline.
    c. Stay away from LOC's.
    d. Stop, Look, Listen, and Smell; take appropriate action.
    e. Move from one concealed area to another.
    f. Use evasion movement techniques.

5. COMMUNICATION AND SIGNALING

    a. Communicate per theater communication procedures, particularly when considering transmitting in the "blind".
    b. Be prepared to use devices on short notice.
    c. Communication/signaling devices may compromise position.

6. RECOVERY OPERATIONS

a. Select site(s) lAW criteria in theater recovery plans.

b. Ensure site is free of hazards; secure personal gear.

c. Select the best area for communications and signaling devices.

d. Observe site for proximity to enemy activity and LOCs.

e. Follow recovery force instructions.

Radio Information (Samples):

PROGRAMMING FILE CHANNEL LIST
FRS - GMRS - PMR - MURS - BUSINESS
MARINE - WEATHER - SAR - HAM - VHF / UHF
FILE NAME: FRS_GMRS_PMR_MURS_BUS_MARINE_WX_HAM_2013F.CSV   INFO: **RADIOFREEQ.WORDPRESS.COM**

MEM CH SLOT	UHF VHF	CHANNEL DESCRIPTION	CHANNEL DISPLAY NAME	FREQUENCY RECEIVE	FREQUENCY TRANSMIT	OFF SET MHZ	PL	TONE HZ	MODE
0	UHF	FRS & GMRS CH 1	FRS 01	462.562500	SIMPLEX	0.0	TX PL	67.0	NFM
1	UHF	FRS & GMRS CH 1	FRS 1	462.562500	SIMPLEX	0.0	TX PL	67.0	NFM
2	UHF	FRS & GMRS CH 2	FRS 2	462.587500	SIMPLEX	0.0	TX PL	67.0	NFM
3	UHF	FRS & GMRS CH 3	FRS 3	462.612500	SIMPLEX	0.0	TX PL	67.0	NFM
4	UHF	FRS & GMRS CH 4	FRS 4	462.637500	SIMPLEX	0.0	TX PL	67.0	NFM
5	UHF	FRS & GMRS CH 5	FRS 5	462.662500	SIMPLEX	0.0	TX PL	67.0	NFM
6	UHF	FRS & GMRS CH 6	FRS 6	462.687500	SIMPLEX	0.0	TX PL	67.0	NFM
7	UHF	FRS & GMRS CH 7	FRS 7	462.712500	SIMPLEX	0.0	TX PL	67.0	NFM
8	UHF	FRS CH 8	FRS 8	467.562500	SIMPLEX	0.0	TX PL	67.0	NFM
9	UHF	FRS CH 9	FRS 9	467.587500	SIMPLEX	0.0	TX PL	67.0	NFM
10	UHF	FRS CH 10	FRS 10	467.612500	SIMPLEX	0.0	TX PL	67.0	NFM
11	UHF	FRS CH 11	FRS 11	467.637500	SIMPLEX	0.0	TX PL	67.0	NFM
12	UHF	FRS CH 12	FRS 12	467.662500	SIMPLEX	0.0	TX PL	67.0	NFM
13	UHF	FRS CH 13	FRS 13	467.687500	SIMPLEX	0.0	TX PL	67.0	NFM
14	UHF	FRS CH 14	FRS 14	467.712500	SIMPLEX	0.0	TX PL	67.0	NFM
15	UHF	GMRS CH 15	GMRS15	462.550000	SIMPLEX	0.0	TX PL	67.0	FM
16	UHF	GMRS CH 16	GMRS16	462.575000	SIMPLEX	0.0	TX PL	67.0	FM
17	UHF	GMRS CH 17	GMRS17	462.600000	SIMPLEX	0.0	TX PL	67.0	FM
18	UHF	GMRS CH 18	GMRS18	462.625000	SIMPLEX	0.0	TX PL	67.0	FM
19	UHF	GMRS CH 19	GMRS19	462.650000	SIMPLEX	0.0	TX PL	67.0	FM
20	UHF	GMRS CH 20	GMRS20	462.675000	SIMPLEX	0.0	TX PL	141.3	FM
21	UHF	GMRS CH 21	GMRS21	462.700000	SIMPLEX	0.0	TX PL	67.0	FM
22	UHF	GMRS CH 22	GMRS22	462.725000	SIMPLEX	0.0	TX PL	67.0	FM
23	UHF	GMRS CH 15 REPEATER 550	GMR15R	462.550000	DUPLEX+	5.0	TX PL	141.3	FM
24	UHF	GMRS CH 16 REPEATER 575	GMR16R	462.575000	DUPLEX+	5.0	TX PL	141.3	FM
25	UHF	GMRS CH 17 REPEATER 600	GMR17R	462.600000	DUPLEX+	5.0	TX PL	141.3	FM
26	UHF	GMRS CH 18 REPEATER 625	GMR18R	462.625000	DUPLEX+	5.0	TX PL	141.3	FM
27	UHF	GMRS CH 19 REPEATER 650	GMR19R	462.650000	DUPLEX+	5.0	TX PL	141.3	FM
28	UHF	GMRS CH 20 REPEATER 675	GMR20R	462.675000	DUPLEX+	5.0	TX PL	141.3	FM
29	UHF	GMRS CH 21 REPEATER 700	GMR21R	462.700000	DUPLEX+	5.0	TX PL	141.3	FM
30	UHF	GMRS CH 22 REPEATER 725	GMR22R	462.725000	DUPLEX+	5.0	TX PL	141.3	FM
31	UHF	PMR446 CH 1	PMR 1	446.006250	SIMPLEX	0.0	TX PL	67.0	NFM
32	UHF	PMR446 CH 2	PMR 2	446.018750	SIMPLEX	0.0	TX PL	67.0	NFM
33	UHF	PMR446 CH 3	PMR 3	446.031250	SIMPLEX	0.0	TX PL	67.0	NFM
34	UHF	PMR446 CH 4	PMR 4	446.043750	SIMPLEX	0.0	TX PL	67.0	NFM
35	UHF	PMR446 CH 5	PMR 5	446.056250	SIMPLEX	0.0	TX PL	67.0	NFM
36	UHF	PMR446 CH 6	PMR 6	446.068750	SIMPLEX	0.0	TX PL	67.0	NFM
37	UHF	PMR446 CH 7	PMR 7	446.081250	SIMPLEX	0.0	TX PL	67.0	NFM
38	UHF	PMR446 CH 8	PMR 8	446.093750	SIMPLEX	0.0	TX PL	67.0	NFM
39	VHF	MURS CH 1	MURS 1	151.820000	SIMPLEX	0.0	TX PL	67.0	NFM
40	VHF	MURS CH 2	MURS 2	151.880000	SIMPLEX	0.0	TX PL	67.0	NFM
41	VHF	MURS CH 3	MURS 3	151.940000	SIMPLEX	0.0	TX PL	67.0	NFM
42	VHF	MURS CH 4 BLUE DOT	MURS 4	154.570000	SIMPLEX	0.0	TX PL	67.0	FM
43	VHF	MURS CH 5 GREEN DOT	MURS 5	154.600000	SIMPLEX	0.0	TX PL	67.0	FM
44	VHF	BUSINESS RED DOT	BUSRED	151.625000	SIMPLEX	0.0	TX PL	67.0	FM
45	VHF	BUSINESS PURPLE DOT	BUSPUR	151.955000	SIMPLEX	0.0	TX PL	67.0	FM

46	VHF	MARINE CH 01A	MAR01A	156.050000	SIMPLEX	0.0	NO PL	00.0	FM
47	VHF	MARINE CH 03A	MAR03A	156.150000	SIMPLEX	0.0	NO PL	00.0	FM
48	VHF	MARINE CH 05A	MAR05A	156.250000	SIMPLEX	0.0	NO PL	00.0	FM
49	VHF	MARINE CH 06	MAR06	156.300000	SIMPLEX	0.0	NO PL	00.0	FM
50	VHF	MARINE CH 07A	MAR07A	156.350000	SIMPLEX	0.0	NO PL	00.0	FM
51	VHF	MARINE CH 08	MAR08	156.400000	SIMPLEX	0.0	NO PL	00.0	FM
52	VHF	MARINE CH 09	MAR09	156.450000	SIMPLEX	0.0	NO PL	00.0	FM
53	VHF	MARINE CH 10	MAR10	156.500000	SIMPLEX	0.0	NO PL	00.0	FM
54	VHF	MARINE CH 11	MAR11	156.550000	SIMPLEX	0.0	NO PL	00.0	FM
55	VHF	MARINE CH 12	MAR12	156.600000	SIMPLEX	0.0	NO PL	00.0	FM
56	VHF	MARINE CH 13	MAR13	156.650000	SIMPLEX	0.0	NO PL	00.0	FM
57	VHF	MARINE CH 14	MAR14	156.700000	SIMPLEX	0.0	NO PL	00.0	FM
58	VHF	MARINE CH 15	MAR15	156.750000	SIMPLEX	0.0	NO PL	00.0	FM
59	VHF	MARINE CH 16 SAFETY	MAR16	156.800000	SIMPLEX	0.0	NO PL	00.0	FM
60	VHF	MARINE CH 17	MAR17	156.850000	SIMPLEX	0.0	NO PL	00.0	FM
61	VHF	MARINE CH 18A	MAR18A	156.900000	SIMPLEX	0.0	NO PL	00.0	FM
62	VHF	MARINE CH 19A	MAR19A	156.950000	SIMPLEX	0.0	NO PL	00.0	FM
63	VHF	MARINE CH 20A	MAR20A	157.000000	SIMPLEX	0.0	NO PL	00.0	FM
64	VHF	MARINE CH 21A	MAR21A	157.050000	SIMPLEX	0.0	NO PL	00.0	FM
65	VHF	MARINE CH 22A	MAR22A	157.100000	SIMPLEX	0.0	NO PL	00.0	FM
66	VHF	MARINE CH 23A	MAR23A	157.150000	SIMPLEX	0.0	NO PL	00.0	FM
67	VHF	MARINE CH 62	MAR62	156.125000	SIMPLEX	0.0	NO PL	00.0	FM
68	VHF	MARINE CH 63A	MAR63A	156.175000	SIMPLEX	0.0	NO PL	00.0	FM
69	VHF	MARINE CH 65A	MAR65A	156.275000	SIMPLEX	0.0	NO PL	00.0	FM
70	VHF	MARINE CH 66A	MAR66A	156.325000	SIMPLEX	0.0	NO PL	00.0	FM
71	VHF	MARINE CH 67	MAR67	156.375000	SIMPLEX	0.0	NO PL	00.0	FM
72	VHF	MARINE CH 68	MAR68	156.425000	SIMPLEX	0.0	NO PL	00.0	FM
73	VHF	MARINE CH 69	MAR69	156.475000	SIMPLEX	0.0	NO PL	00.0	FM
74	VHF	MARINE CH 71	MAR71	156.575000	SIMPLEX	0.0	NO PL	00.0	FM
75	VHF	MARINE CH 72	MAR72	156.625000	SIMPLEX	0.0	NO PL	00.0	FM
76	VHF	MARINE CH 73	MAR73	156.675000	SIMPLEX	0.0	NO PL	00.0	FM
77	VHF	MARINE CH 74	MAR74	156.725000	SIMPLEX	0.0	NO PL	00.0	FM
78	VHF	MARINE CH 75	MAR75	156.775000	SIMPLEX	0.0	NO PL	00.0	FM
79	VHF	MARINE CH 76	MAR76	156.825000	SIMPLEX	0.0	NO PL	00.0	FM
80	VHF	MARINE CH 77	MAR77	156.875000	SIMPLEX	0.0	NO PL	00.0	FM
81	VHF	MARINE CH 78A	MAR78A	156.925000	SIMPLEX	0.0	NO PL	00.0	FM
82	VHF	MARINE CH 79A	MAR79A	156.975000	SIMPLEX	0.0	NO PL	00.0	FM
83	VHF	MARINE CH 80A	MAR80A	157.025000	SIMPLEX	0.0	NO PL	00.0	FM
84	VHF	MARINE CH 81A	MAR81A	157.075000	SIMPLEX	0.0	NO PL	00.0	FM
85	VHF	MARINE CH 88A	MAR88A	157.425000	SIMPLEX	0.0	NO PL	00.0	FM
86	VHF	WEATHER NOAA CH WX 1	WX 1	162.550000	RX ONLY	0.0	NO PL	00.0	FM
87	VHF	WEATHER NOAA CH WX 2	WX 2	162.400000	RX ONLY	0.0	NO PL	00.0	FM
88	VHF	WEATHER NOAA CH WX 3	WX 3	162.475000	RX ONLY	0.0	NO PL	00.0	FM
89	VHF	WEATHER NOAA CH WX 4	WX 4	162.425000	RX ONLY	0.0	NO PL	00.0	FM
90	VHF	WEATHER NOAA CH WX 5	WX 5	162.450000	RX ONLY	0.0	NO PL	00.0	FM
91	VHF	WEATHER NOAA CH WX 6	WX 6	162.500000	RX ONLY	0.0	NO PL	00.0	FM
92	VHF	WEATHER NOAA CH WX 7	WX 7	162.525000	RX ONLY	0.0	NO PL	00.0	FM
93	VHF	SEARCH RESCUE EMT	SAREMT	155.160000	SIMPLEX	0.0	TX PL	127.3	FM
94	VHF	HAM 2 METER 146.42	HAM 42	146.420000	SIMPLEX	0.0	TX PL	100.0	FM
95	VHF	HAM 2 METER 146.52	HAM 52	146.520000	SIMPLEX	0.0	TX PL	100.0	FM
96	VHF	HAM 2 METER 146.55	HAM 55	146.550000	SIMPLEX	0.0	TX PL	100.0	FM
97	UHF	HAM 446.0	HAM	446.000000	SIMPLEX	0.0	TX PL	100.0	FM
98	UHF	HAM 446.03	HAM U3	446.030000	SIMPLEX	0.0	TX PL	100.0	FM

WALLET SIZE

PREPPER & SURVIVALIST SHTF FREQUENCIES
2-WAY RADIO COMMUNICATIONS

RADIO SERVICE	CHANNEL NAME	FREQUENCY MHZ	DESCRIPTION MODE
FRS UHF	FRS 3	462.6125 FM	PREPPER
GMRS UHF	GMRS17	462.6000 FM	SURVIVALIST
GMRS UHF	GMRS20	462.675+ FM	PL141.3RPTR+5MHz
PMR UHF	PMR 3	446.03125FM	PREPPER
MURS VHF	MURS 3	151.940 FM	PREPPER
CB AM	CB 3AM	26.985 AM	PREPPER
CB AM	CB 9AM	27.065 AM	HIGHWAY SAFETY
CB SSB	CB 36U	27.365 USB	SURVIVALIST
CB SSB	CB 37U	27.375 USB	PREPPER
CB FREEBAND	FB368U	27.368 USB	SURVIVALIST
CB FREEBAND	FB378U	27.378 USB	PREPPER
CB FREEBAND	FB425U	27.425 USB	SURVIVALIST
LOWBAND VHF	LOW334	33.400 FM	SURVIVALIST
HAM UHF	HAM U3	446.030 FM	PREPPER
HAM VHF	HAM 52	146.420 FM	PREPPER
HAM VHF	HAM 52	146.520 FM	HAM CALLING
HAM VHF	HAM 55	146.550 FM	SURVIVALIST
HAM HF	HAM10M	28.305 USB	PREPPER
HAM HF	HAM20M	14.242 USB	PREPPER
HAM HF	HAM40M	7.242 LSB	PREPPER NETS
HAM HF	HAM60M	5.357 USB	SURVIVALIST NVIS
HAM HF	HAM80M	3.818 LSB	PREPPER NETS
LAND SAR VHF	SAREMT	155.160 FM	SEARCH&RESCUE
MARINE VHF	MAR 16	156.800 FM	SAFETY CALLING
MARINE VHF	MAR 72	156.625 FM	BOAT PREPPER

3-3-3
RADIO PLAN
CHANNEL 3
EVERY 3 HOURS
FOR 3 MINUTES
More frequencies at: RADIOFREEQ.WORDPRESS.COM

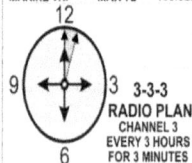

SHTF FREQUENCY LIST

THIS CHART OF ACTIVE SURVIVALIST AND PREPPER FREQUENCIES WAS RELEASED INTO THE PUBLIC DOMAIN IN 2013 BY RADIOMASTER REPORTS.

More information about this list, and frequency charts with programming files are available at RADIOFREEQ.WORDPRESS.COM

3-3-3 Radio Plan

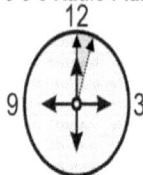

For SHTF Communications. Turn on your radio.

Every 3 hours. For 3 minutes. Channel 3.

3-3-3 RADIO PLAN - The Survivalist Radio Schedule
This is the "When, Where, and How" to make radio contact with each other for SHTF. The 333 Radio Plan was designed for SHTF communications, when normal methods of communication fail. Versions of the 3-3-3 are used by survivalist, prepper, and emergency communications groups worldwide. It is based on the easy-to-remember "Survival Rule of Threes". It is often called an emergency radio schedule or sked.

DESK SIZE

PREPPER & SURVIVALIST SHTF FREQUENCIES
2-WAY RADIO COMMUNICATIONS

RADIO SERVICE	CHANNEL NAME	FREQUENCY MHZ	DESCRIPTION MODE
FRS UHF	FRS 3	462.6125 FM	PREPPER
GMRS UHF	GMRS17	462.600 FM	SURVIVALIST
GMRS UHF	GMRS20	462.675+ FM	PL141.3RPTR+5MHz
PMR UHF	PMR 3	446.03125FM	PREPPER
MURS VHF	MURS 3	151.940 FM	PREPPER
CB AM	CB 3AM	26.985 AM	PREPPER
CB AM	CB 9AM	27.065 AM	HIGHWAY SAFETY
CB SSB	CB 36U	27.365 USB	SURVIVALIST
CB SSB	CB 37U	27.375 USB	PREPPER
CB FREEBAND	FB368U	27.368 USB	SURVIVALIST
CB FREEBAND	FB378U	27.378 USB	PREPPER
CB FREEBAND	FB425U	27.425 USB	SURVIVALIST
LOWBAND VHF	LOW334	33.400 FM	SURVIVALIST
HAM UHF	HAM U3	446.030 FM	PREPPER
HAM VHF	HAM 52	146.420 FM	PREPPER
HAM VHF	HAM 52	146.520 FM	HAM CALLING
HAM VHF	HAM 55	146.550 FM	SURVIVALIST
HAM HF	HAM10M	28.305 USB	PREPPER
HAM HF	HAM20M	14.242 USB	PREPPER
HAM HF	HAM40M	7.242 LSB	PREPPER NETS
HAM HF	HAM60M	5.357 USB	SURVIVALIST NVIS
HAM HF	HAM80M	3.818 LSB	PREPPER NETS
LAND SAR VHF	SAREMT	155.160 FM	SEARCH&RESCUE
MARINE VHF	MAR 16	156.800 FM	SAFETY CALLING
MARINE VHF	MAR 72	156.625 FM	BOAT PREPPER

VERSION: SHTF FREQ LIST 2013E

3-3-3
RADIO PLAN
CHANNEL 3
EVERY 3 HOURS
FOR 3 MINUTES

2013 Public Domain
Source:
RADIOMASTER
REPORTS

More frequencies at: RADIOFREEQ.WORDPRESS.COM

ABOUT THE 3-3-3 RADIO PLAN
Here's how the 3-3-3 Radio Plan works:
Turn on your radio. Every 3 hours. For 3 minutes. Channel 3.

WHEN: EVERY 3 HOURS
Always use your Local Time for local area communications with the 3-3-3 Radio Plan. At the "top of the hour", each 3 hours:
Noon, 3pm, 6pm, 9pm.
Midnight, 3am, 6am, 9am.

HOW LONG: FOR 3 MINUTES
At the top of every 3rd hour, turn on your radio. Even if you don't need to make a call yourself, always turn on your radio and listen for calls for at least 3 minutes. This is because you never know if someone may be trying to reach you, or may need help. If you need to check in, make a short transmission at this time. Say "This is me, just checking in." If you have sufficient battery power, or if you have not connected in for a while, then you should listen for 15 minutes.

ACCURATE TIME KEEPING
Synchronize your watch with others whenever possible. If you doubt your watch accuracy, compensate by keeping your radio on for a longer duration, before and after every 3rd hour. If you don't have a watch, try listening to an AM broadcast radio station, they always identify their call letters at the top of each hour.

WHERE: CHANNEL 3
Channel 3 usually applies to CB, FRS, or MURS. These are the most common types of radios used. If your group has a different designated SHTF channel or Prepper SHTF HAM frequency, you should use it instead of Channel 3. For example, the ham 2 meter simplex calling channel 146.520 MHz. The rest of the 3-3-3 Radio Plan remains the same. Keep it simple.

HOW IT WORKS: FEATURES OF THE 3-3-3 RADIO PLAN

1. Easy for everyone to remember the "Rule of Three".
2. Conserves precious battery life for walkie talkies.
3. Gets everyone on the air at the same time.
4. Sets a schedule of 8 times per day to call each other.
5. Avoids impractical hourly schedules.
6. Enables the use of short transmissions for optimum success and security.
7. Three hours is enough time to rest in a survival situation.
8. A person can walk 8 miles in 3 hours, the practical distance limit of handheld radios over average terrain.

VERSION: SHTF_FREQUENCY_LIST_2013E

Glossary - Terms, Acronyms, Abbreviations, and a few limited Definitions

ACE
Ammo, Casualties, Equipment

AO
Area of Operations

Buddy Team
Two (or three) inseparable battle buddies. Implemented for accountability, buddy team rushes/maneuvers, and wellbeing of a group.

CM
Citizen Manual

CPR
Cardiopulmonary Resuscitation

CSS
Combat Service Support

DAR
Designated Area for Recovery

EMT
Emergency Medical Technician

FM
Field Manual

GTX
Gore-Tex

IFAK

Individual First Aid Kit

Kilometer
1,000 Meters

LBE
Load Bearing Equipment

MAG
Mutual Assistance Group

NPA
Nasopharyngeal Airway

OPFOR
Opposing Force

PMA
Positive Mental Attitude

QRF
Quick Reaction Force

SAFE
Selected Area for Evasion

SOP
Standard Operation Procedure

TCCC
Tactical Combat Casualty Care

TQ
Tourniquet

WB

Waterproof/Breathable

W/P
Waterproof

WFA
Wilderness First Aid

WFR
Wilderness First Responder

References:

-Extreme Alpinism by Mark F. Twight
-Mountaineering, The Freedom of the Hills by The Mountaineers
-The Professional Citizen Project CM1 by Jack Morris
-Outdoor Life Ultimate Bushcraft Survival Manual, Outdoor Life
-Mountain Warfare and Cold Weather Operations, US Army FM
-Jungle Operations, US Army/USMC FM
-Desert Operations, US Army/USMC FM
-Infantry Small-Unit Mountain Operations, US Army FM
-S.E.R.E. Student Handbook USN & USMC FM
-Radio Master Reports
-Citizen Manual 1 (CM-1) Individual Tactical Skills - Jack Morris

www.ingramcontent.com/pod-product-compliance
Lightning Source LLC
Chambersburg PA
CBHW070309200326
41518CB00010B/1951